Thomas Werner Degen

Portable Devices for Mobile Health Monitoring

Thomas Werner Degen

Portable Devices for Mobile Health Monitoring

Ubiquitous Amplifiers for Bioelectric Events

Südwestdeutscher Verlag für Hochschulschriften

Impressum/Imprint (nur für Deutschland/only for Germany)
Bibliografische Information der Deutschen Nationalbibliothek: Die Deutsche Nationalbibliothek verzeichnet diese Publikation in der Deutschen Nationalbibliografie; detaillierte bibliografische Daten sind im Internet über http://dnb.d-nb.de abrufbar.
Alle in diesem Buch genannten Marken und Produktnamen unterliegen warenzeichen-, marken- oder patentrechtlichem Schutz bzw. sind Warenzeichen oder eingetragene Warenzeichen der jeweiligen Inhaber. Die Wiedergabe von Marken, Produktnamen, Gebrauchsnamen, Handelsnamen, Warenbezeichnungen u.s.w. in diesem Werk berechtigt auch ohne besondere Kennzeichnung nicht zu der Annahme, dass solche Namen im Sinne der Warenzeichen- und Markenschutzgesetzgebung als frei zu betrachten wären und daher von jedermann benutzt werden dürften.

Coverbild: www.ingimage.com

Verlag: Südwestdeutscher Verlag für Hochschulschriften GmbH & Co. KG
Heinrich-Böcking-Str. 6-8, 66121 Saarbrücken, Deutschland
Telefon +49 681 37 20 271-1, Telefax +49 681 37 20 271-0
Email: info@svh-verlag.de

Approved by: Zürich, ETHZ, Diss., 2011

Herstellung in Deutschland:
Schaltungsdienst Lange o.H.G., Berlin
Books on Demand GmbH, Norderstedt
Reha GmbH, Saarbrücken
Amazon Distribution GmbH, Leipzig
ISBN: 978-3-8381-3193-1

Imprint (only for USA, GB)
Bibliographic information published by the Deutsche Nationalbibliothek: The Deutsche Nationalbibliothek lists this publication in the Deutsche Nationalbibliografie; detailed bibliographic data are available in the Internet at http://dnb.d-nb.de.
Any brand names and product names mentioned in this book are subject to trademark, brand or patent protection and are trademarks or registered trademarks of their respective holders. The use of brand names, product names, common names, trade names, product descriptions etc. even without a particular marking in this works is in no way to be construed to mean that such names may be regarded as unrestricted in respect of trademark and brand protection legislation and could thus be used by anyone.

Cover image: www.ingimage.com

Publisher: Südwestdeutscher Verlag für Hochschulschriften GmbH & Co. KG
Heinrich-Böcking-Str. 6-8, 66121 Saarbrücken, Germany
Phone +49 681 37 20 271-1, Fax +49 681 37 20 271-0
Email: info@svh-verlag.de

Printed in the U.S.A.
Printed in the U.K. by (see last page)
ISBN: 978-3-8381-3193-1

Copyright © 2012 by the author and Südwestdeutscher Verlag für Hochschulschriften GmbH & Co. KG and licensors
All rights reserved. Saarbrücken 2012

Portable Devices
for
Mobile Health Monitoring

Thomas Werner Degen

February 28, 2012

To Isabelle and Raphael, to Christina
and to my whole family
for all the hours spent
away from you

> Ça me prend,
> parfois,
> au coucher du soleil:
> l'impression de vieillir
> dans une salle d'attente
>
> *Franco de Guglielmo*

CONTENTS

Abstract 1

Zusammenfassung (German) 3

1. Introduction 5
   1.1 Motivation and Objectives 5
   1.2 Structure of the Thesis 6

2. Overview of the Research Field 7
   2.1 Signal Description 8
   2.2 Signal Acquisition 9
      2.2.1 Electrodes 9
      2.2.2 Artifacts 13
      2.2.3 Patient Isolation and Common Mode Rejection Ratio 14
      2.2.4 Common Mode to Differential Mode Conversion .. 19
      2.2.5 Total Common Mode Rejection Ratio 25
      2.2.6 Reducing the power-line Interference by Filtering . 28
   2.3 Amplifiers for Bioelectric Events 29
      2.3.1 Amplifiers using Tripolar Electrodes 30
      2.3.2 Amplifiers for Two Electrodes 30
      2.3.3 Amplifiers using Reference Electrodes 32
      2.3.4 AC-coupled Amplifiers 37
      2.3.5 DC-coupled Amplifiers 42
      2.3.6 Review of Amplifiers for Bioelectric Events 45
   2.4 Figures of Merit 47

3. Monitoring Electrode-Skin Impedance Mismatch 49
    3.1 Problem Statement 49
    3.2 Prior Art 49
    3.3 Method 50
           3.3.1 Experimental Results 53
    3.4 Novelty 59

4. Low-Noise Two-wired Buffer Electrodes 61
    4.1 Problem Statement 61
    4.2 Prior Art 61
           4.2.1 Pilot Study 62
    4.3 Method 64
           4.3.1 Results 69
    4.4 Novelty 72

5. Gain Adaptation for Amplifying Electrodes 73
    5.1 Problem Statement 73
    5.2 Prior Art 73
    5.3 Method 74
           5.3.1 Implementation 78
           5.3.2 Results 82
    5.4 Novelty 87

6. Two-wired Amplifying Electrodes 89
    6.1 Problem Statement 89
    6.2 Prior Art 90
    6.3 Method 90
           6.3.1 Two-Wired Amplifying Electrodes 91
           6.3.2 Amplifier Stage 94
           6.3.3 Results 105
    6.4 Evaluation of Additional Op-Amps 117
    6.5 Novelty 119

7. Conclusion 121

Appendix   127

A. Figures of Merit 129

B. Schematics for Simulation 131
    B.1 Two-Wired Amplifying Electrode 131
    B.2 Low-Voltage Two-Wired Amplifying Electrode 135
    B.3 Two-Wired Buffer Electrodes 138

Bibliography 141

Acknowledgments 165

Index 167

ABSTRACT

In this thesis we investigate methods to improve amplifiers for bioelectric events such as ECG (electrocardiogram) and EEG (electroencephalogram). The focus being on noise-optimization, power minimization, wearability and the reduction of motion artifacts. The research was motivated by the challenges arising while building wearable medical devices. The main topic is active electrodes and their application.

The most important results were described in four publications (IEEE Transactions on Biomedical Engineering).

These include:

- The measurement of the electrode-skin impedance mismatch between two electrodes while concurrently measuring a bioelectrical signal without degradation of the performance of the amplifier.

- The efficient, noise-optimized measurement of bioelectrical signals by means of two-wired active buffer electrodes.

- The reduction of power-line interference when using amplifying electrodes by means of autonomous adaption of the gain of the subsequent differential amplification.

- The design of an amplifier with two-wired amplifying electrodes having a gain of 40 dB. The amplifier's features include offset compensation, CMRR improvement in software and a bandwidth extending down to DC.

These results contribute to the development of the next generation of wearable and highly integrated medical monitoring devices suited for daily use.

ZUSAMMENFASSUNG

Im Rahmen dieser Dissertation befassen wir uns mit Methoden bioelektrische Verstärker zu verbessern. Der Schwerpunkt liegt auf Verstärkern für EKG (Elektrokardiogramm) und EEG (Elektroenzephalogramm). Besonderes Gewicht liegt auf Rauschoptimierung, geringem Verbrauch, Tragbarkeit und Reduktion der Bewegungs-Artefakte. Die Forschung wurde motiviert durch die bei der Entwicklung tragbarer biomedizinischer Verstärker auftretenden Fragestellungen. Aktive Elektroden und deren Anwendung bilden einen Schwerpunktbereich.

Die wichtigsten Resultate dieser Arbeit wurden in vier Publikationen veröffentlicht (IEEE Transactions on Biomedical Engineering).

Die Resultate beinhalten:

- Die Messung des Ungleichgewichts der Elektroden-Haut Übergangsimpedanz bei gleichzeitiger störungsfreier Aufzeichnung eines bioelektrischen Signals.

- Die rauscharme und leistungseffiziente Messung von bioelektrischen Signalen mittels zweiadriger aktiver Puffer-Elektroden.

- Die Reduktion der 50 Hz Interferenz bei der Verwendung verstärkender Elektroden durch die autonome Anpassung der nachfolgenden differentiellen Verstärkung.

- Die Entwicklung eines Verstärkers mit zweiadrigen verstärkenden Elektroden mit einer Verstärkung von 40 dB. Die Eigenschaften des Verstärkers umfassen Offset-Kompensation, Software-basierte Wiederherstellung der CMRR und eine Bandbreite bis und mit DC.

Diese Resultate tragen bei zur Entwicklung der nächsten Generation von tragbaren und hoch-integrierten medizinischen Geräten für den täglichen Einsatz.

1. INTRODUCTION

1.1 Motivation and Objectives

IN spite of more powerful examination techniques such as computer tomography, EEG and ECG are still very widely used. EEG is the only medical indication to diagnose an ongoing epileptic seizure with certitude. While ECG is, to the author's knowledge, still the most power-efficient method to monitor the heart beat. Both EEG and ECG have an excellent temporal resolution but a poor spatial resolution.

There is a regained interest in ECG and EEG driven by a new paradigm. Instead of collecting a single data point (e.g., the blood pressure of a person at the time of a visit to the doctor) it is better to monitor the variation of several parameters during the day or the progression of a measurement over a long period. This allows for a better evaluation of an individual's health condition and even helps to foresee a critical condition before it occurs. This is a concept which has long been well accepted in the field of industrial maintenance but has yet to become standard for human health.

The new paradigm led to the development of new portable monitoring devices intended for daily use. Working in this field we realized that there is little knowledge about how to miniaturize an ECG or EEG measurement system while maintaining the clinical use for the physician.

Body-worn devices are also very susceptible to a variation of the contact between the skin and the electrodes. As a result of the electrodes being pressed to the skin large variations of the baseline[1] appear for example during physical activity. These variations are a result of the variation of the contact pressure.

The methods developed during this thesis serve the purpose of enhancing wearable bioelectric monitoring devices.

---

[1] the baseline corresponds to the period of zero muscle activity of the heart and normally is flat

1.2 Structure of the Thesis

The thesis is divided into three parts. In the first part we give a short recapitulation of the theory behind biopotential measurement systems. We start with the most important components of a bioelectric measurement chain and then discuss the different amplifier topologies.

The second part consists of a synopsis of the four publications written during this thesis. Each chapter is divided into four parts, namely problem statement, prior art, method and novelty.

The last part contains the conclusion followed by the appendix. In the appendix there is a recapitulation of the figures of merit as well as the simulation schematics of the circuits described in the second part.

2. OVERVIEW OF THE RESEARCH FIELD

THE theory of bioelectrical measurements is well established and many excellent books have been written. In the bibliography (chapter B.3) at the end of this report we compiled a list of publications covering the design of electrodes and amplifiers. For a detailed overview we recommend the following books: 'Handbuch Medizintechnik' [*Menk* 89], 'Medical Instrumentation, Applications and Design' [*Webs* 98] and 'Klinische Elektroenzephalographie' [*Zsch* 02].

The objective of this chapter is to give a broad overview of the general theory required as background for the more specific problems treated in the second part of this thesis. The emphasis is placed on the electrical characteristics of the bioelectric signals as well as the important design parameters. The design parameters will be summarized at the end of this chapter in section 2.4.

In this thesis we will analyze the measurement situation from the point of view of an electrical engineer. Clinical interpretation and biochemical processes are not within the scope of this overview.

We start with a short overview of the bioelectric signal chain:

At the origin of a bioelectric measurement is always the source. For an ECG the source is the muscle contraction of the heart. To measure a signal from such a source we then need a transducer which, in the case of bio-electric measurements, is a set of electrodes. The bioelectric signal is then amplified and presented to the user in one form or another. The amplified signal can also be stored for future reference.

The bioelectric signal is a current or voltage where both reference points are at a body-derived potential. The body itself is connected capacitively to the *earth*, e.g., the electric potential of the surface of the earth. The amplifier may have its own power supply, e.g., batteries and therefore have its one reference voltage which we call *system ground*. The relation between the system ground and the earth will be treated in more detail in section 2.2.3

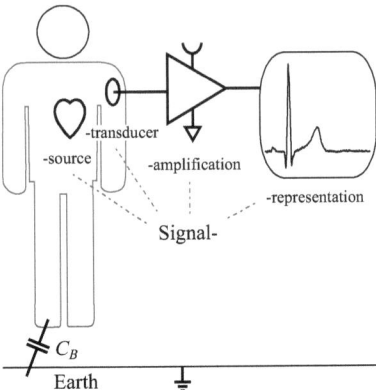

Fig. 2.1: Bioelectrical signal chain consisting of the signal-source, -transducer, -amplifier and -representation.

2.1 Signal Description

ECG and EEG are the two most commonly measured biopotentials. Table 2.1 resumes the most important signal descriptions for ECG and EEG.

Tab. 2.1: Typical values for ECG and EEG

	Amplitude[a]	Bandwidth[b]	f_0^c	Noise[d]
ECG	1 mV	0.1 - 200 Hz	0.016 Hz	1.4 μV_{rms}
EEG	10-100 μV	10 - 1000 Hz	1 Hz	1.5 μV_{rms}

[a] typical peak amplitude, depends on clinical application and location of the electrodes [Zsch 02], [Menk 89]
[b] typical bandwidth, depends on clinical application [Zsch 02], [Menk 89]
[c] recommended corner frequency of the highpass filter used to remove the electrode-electrolyte offset voltage [AAMI 99]
[d] typical voltage noise of electrodes when using Ag/AgCl-electrodes integrated over the respective signal bandwidth

The value of the integrated voltage noise of the electrode was introduced in table 2.1 to give a rough estimate of the lowest detectable bioelectric signal. The origin of the presented value is detailed in section 2.2.1. The implication of the electrode's voltage noise on amplifier design is discussed in more detail in section 2.3.4 and 2.3.5.

2.2 Signal Acquisition

The main difference between an amplifier for an arbitrary electric signal and an amplifier for bioelectric events is the specific requirement for patient safety. Safety regulations demand the patient be isolated from any electric path to the earth. In addition to the electric isolation, the bioelectric signal has a relatively large source resistance (i.e., several kΩ) and the electrode-skin interface may vary considerably from person to person. Even between different electrodes on the same patient the electrode-skin impedance as well as the electrode-electrolyte offset voltage can vary considerably as shown in the following section.

We now resume the specific elements in the signal acquisition chain of bioelectric measurement systems starting with the electrodes.

2.2.1 Electrodes

Electrodes are the transducers between the body and the electro-physiological signal measured by a bioelectrical amplifier.

It is important to understand that there is a change in the nature of the electric current taking place at the electrode site. Inside the skin the electric current is transported by ions, whereas from the electrode onwards, the current is carried on by free electrons inside the metal. As a consequence, there is an electro-chemical reaction taking place at the site of the electrode, often supported by the electrolyte [Menk 89]. This leads to some important conclusions:

- The material of the electrode directly impacts the quality of the electro-chemical reaction. The best material in terms of noise is Ag/AgCl (silver/silver chloride) [Huig 02, Fern 00, Godi 91]
- After applying an electrode to the skin, the electrode-skin interface requires a time of about 15 minutes to build an electro-chemical equilibrium. During this time excessive noise is measured [Huig 02].

Electrode-Noise

The resolution in bioelectric measurements is limited by the thermal noise of the electrodes. A measurement of the spectral voltage-noise density for the two most commonly used electrodes is drawn in Fig. 2.2.

In Fig. 2.2 the measured spectral voltage-noise density of the electrodes is compared to the theoretical thermal voltage noise density of a resistor

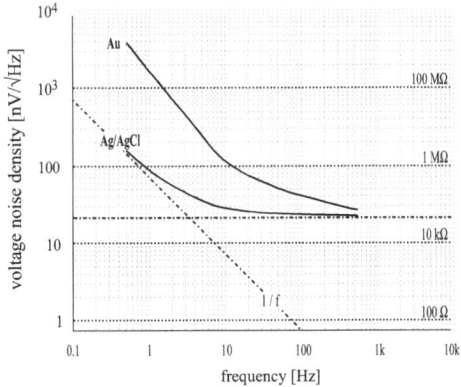

Fig. 2.2: Typical measured spectral voltage-noise density of Ag/AgCl and Au electrodes compared to the theoretical thermal noise of a resistor. (*The measurements of the electrode noise were carried out by biosemi [bios], explanations at "http://www.biosemi.com/faq/without_paste.html"*)

calculated in function of its resistance and temperature:

$$\overline{v}_{nR} \approx \sqrt{4k_B T R} \qquad (2.1)$$

Where k_B stands for the Boltzmann constant, T for the temperature in Kelvin and R for the resistance of the electrode.

From Fig. 2.2 we can estimate the integrated spectral voltage-noise density of an Ag/AgCl electrode over the bandwidth of 10 Hz to 1 kHz to about 1.4 μV_{rms}. This was used for the electrode noise reported in table 2.1 for an ECG recording.

The integrated spectral voltage-noise density is only a rough measurement because it holds no information on the contribution at a specific frequency. A closer look at the spectral voltage-noise density of an Ag/AgCl electrode reveals a resistive behavior for frequencies above about 3 Hz while the density varies inversely proportional to f for lower frequencies. Above 3 Hz the inherent noise of an Ag/AgCl electrode corresponds to the thermal noise of a 25 kΩ resistor (estimated from Fig. 2.2). This supports the often encountered recommendation to replace electrodes when the corresponding electrode-skin impedance (often measured at 10 Hz) raises over 20 kΩ.

For the design of an amplifier for bioelectric events the spectral voltage-noise density of the electrodes yields the target value for the input-referred spectral voltage-noise density of the amplifier (see also section 2.3.4 and 2.3.5).

Electrode-Skin Interface

A simple model for the electrode-skin interface which is widely used in literature is shown in Fig. 2.3. The values for this model are extracted from various sources as for example [Yama 77] and [Mett 90]. The model shows typical values for an Ag/AgCl electrode with good ohmic contact such as the commonly used disposable pre-gelled electrodes. If a more complex model of the electrode-skin interface is required we refer to the extensive work of Neumann in [Webs 98].

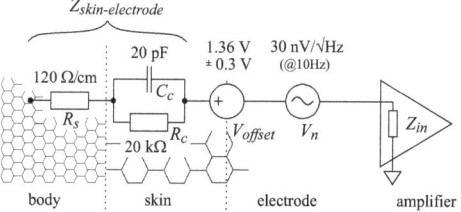

Fig. 2.3: A simplified model of the electrode-skin interface for a pre-gelled Ag/AgCl electrode.

To the offset voltage V_{offset} is added in series the voltage noise source V_n which corresponds to the inherent noise of the electrode described in the previous section. As an example, the spectral voltage-noise density for an Ag/AgCl electrode is depicted in Fig. 2.2.

For different dry electrodes, the typical value for the resistor R_c is much higher, i.e., in the range of 1.4 MΩ [Burk 00]. The same is true for integrated electrodes, which have much higher source resistances (not shown in figure 2.3) due to the absence of Ag/AgCl or similar suitable materials [Tahe 94].

In general, the skin impedance decreases with frequency and depends on size, material and pressure of the applied electrode. A set of measurements is for example reported in [Rose 88].

Between the electrodes and the skin there is always a supplementary layer. Either gel is applied manually or transpiration will accumulate and form an electrolyte between the skin and electrode. This additional interface will improve the ohmic contact by increasing the contact surface. But this interface is a half-cell structure (interface between metal and electrolyte) resulting in a redox reaction. This reaction pumps electric charges between electrode and the electrolyte, resulting in a potential difference V_{offset} between the electrode and the electrolyte of typically 1.36 V (this is the standard electrode potential involved in the redox reaction $Cl_2 + 2e- \leftrightarrow 2Cl-$). The electrode potential varies with temperature, pressure and, most important, concentration of the electrolyte. As a result, the voltage across the electrode-electrolyte interface varies between different electrodes. In bioelectric applications, a variation of up to ±300 mV is to be expected [*AAMI* 99]. A detailed description is found in the fifth chapter of [*Webs* 98].

As a consequence of the unknown offset voltage V_{offset} between electrode and skin, there is no method to measure the electrode-skin resistance (i.e., the electrode-skin impedance at DC). In most applications the electrode-skin impedance is measured at around 10 Hz.

The electrode-electrolyte interface will also contribute to the voltage change resulting from electrode movement, an effect which is part of the noise summarized by motion artifacts (see section 2.2.2).

Aside from the commonly used conductive electrodes there is a second group of electrodes, called polarizable electrodes.

Polarizable Electrodes

Polarizable electrodes are electrodes which do not actually conduct an electron current. Instead, there is a displacement current (a change of the local E-field). All capacitive electrodes are polarizable electrodes. The name *polarizable electrodes* comes from the fact the electrode becomes polarized when a voltage is applied across it.

By making electrodes out of rubber mixed with graphite, the interface comes close to the model of a perfectly polarizable electrode. Polarizable electrodes do not have a half-cell voltage because no redox reaction is taking place [*Zsch* 02].

Polarizable electrodes cannot be used with DC-coupled amplifiers, they must be used with AC-coupled amplifiers, which are discussed in section 2.3.4. Polarizable electrodes do not allow measuring the DC value of a bioelectric signal.

In theory capacitors are noiseless. Therefore we should think that polarizable electrodes are the perfect choice for bioelectric signals. Although ideal capacitors are noiseless, they shape the noise of the resistors in the circuit and therefore the noise of polarizable electrodes cannot be discussed without taking into account the input stage of the amplifier for bioelectric events used, e.g., its input impedance. The noise of AC-coupled input stages is discussed in more detail in section 2.3.4.

Polarizable electrodes themselves are not noiseless because of dielectric noise (skin) and motion artifacts. Motion artifacts will be discussed in section 2.2.2.

Active Electrodes

Electrodes comprising electronics to amplify the bioelectric signal are called *active electrodes*. The amplification does not necessarily have to be a voltage amplification. In fact, most commercially active electrodes provide a voltage gain of one and are therefore called *buffer electrodes*. Buffer electrodes provide, however, an impedance transformation by reducing the source impedance seen by the remote bioelectrical amplifier. Most buffer electrodes described in literature employ an op-amp (operational amplifier) in the voltage-follower configuration [Ko 98].

As a general rule it can be said that active electrodes reduce the negative effects of long wires (mainly capacitive interference) but require additional leads for their power supply.

The number of wires can be the limiting factor for some applications such as a 128-lead or 256-lead EEG. In addition, large number of wires per electrode increases the total wire thickness of the electrode and therefore its stiffness. This may lead to augmented motion artifacts and increases the size of the connectors which is a limiting factor for wearable devices.

Active electrodes are an important part of this work. We will describe two-wired active buffer electrodes in chapter 4 and two-wired amplifying electrodes in chapter 6.

2.2.2 Artifacts

Any part of the recorded signal which has not its origin in the bioelectric effect under observation is called an artifact. This can be other bioelectric signals like muscle activity which is recorded as part of an EEG. It can also be the result of an electromagnetic interference from various origins (e.g., power-line interference) or being a result of movement of the electrodes (i.e., motion artifacts). A good overview of commonly encountered artifacts can be found in [Webs 84] and [Mett 90].

To reduce artifacts some basic rules should always be followed:

- In order to reduce magnetic coupling of the power-line voltage to the system, long wires should be twisted whenever possible to minimize loops.

- To reduce capacitive coupling of the power-line interference wires should be driven with the lowest possible impedance and, if possible, shielded with a low impedance potential which is close to the body voltage (guarding).

- In the case of shielded wires their movement should be reduced as much as possible as to reduce tribo-electric noise generated by friction and deformation of the insulation.
- Wires connecting to the electrodes should be as flexible as possible in order to reduce motion artifacts.

Motion Artifacts

One of the most severe limitations for body-worn bioelectrical amplifiers is motion artifacts, i.e., signal disturbances due to motion of the wires, the electrodes or the subject. It has been demonstrated that motion artifacts in general scale inversely to the input resistance of the amplifier for bioelectric events up to an input resistance of 1 GΩ [Zipp 79b]. Motion artifacts do also scale linearly with the input bias current, which should be kept below 50 pA [Zipp 79b].

To reduce motion artifacts by electrical means several methods have been proposed: Some motion artifacts are not in the same frequency band as the bioelectric signal and can be filtered out.

The most effective methods are those which reduce the motion artifact at the source: skin-abrasion or the use of micromachined electrodes [Gris 02]. Unfortunately, these methods are not well suited for long-term monitoring applications as the skin grows back and micromachined electrodes may lead to skin irritation.

Some authors use capacitive coupled electrodes to enhance immunity to motion artifacts [Burk 00] and [Harl 03]. Yet, the claim that capacitive coupled electrodes reduce motion artifact has never been proven.

Capacitive electrodes (i.e., polarizable electrodes) suffer from an additional kind of motion artifacts which results from motion-related changes of their coupling capacity. Motion artifacts of polarizable electrodes are discussed in section 2.3.4.

2.2.3 Patient Isolation and Common Mode Rejection Ratio

The standard for patient isolation requires that the total current flowing through the patient and the measurement equipment to earth does not exceed a peak value of 50 μA_{rms}, even in the unlikely event that the patient touches a power outlet. These and other requirements are regulated by the AAMI standard (Association for the Advancement of Medical Instrumentation) [*AAMI* 99] and the IEC (International Electrotechnical Commission) standard [*IEC* 08]. As a result, the signal

2.2. Signal Acquisition

source, e.g., the human body, cannot be grounded (connected by a low impedance path to earth) as is the case for most electric measurements, but must be kept electrically isolated. Equally, the measurement equipment must provide some means of isolation, this can be achieved by using capacitive electrodes, by using an isolated battery supply (e.g., wearable devices) or by using isolation amplifiers.

A typical measurement situation is shown in Fig. 2.4 with a three-lead ECG as an example for an amplifier for bioelectric events.

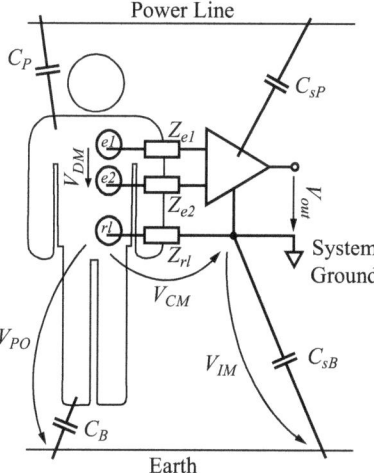

Fig. 2.4: A typical measurement situation for a three-lead ECG including power-line interference.

As in bioelectric measurements the human body is floating, this implies that there is only a capacitive coupling of the patient to both the earth (C_B) and the power mains (C_P). The capacitive coupling from the body to the earth is usually stronger than the capacitive coupling to the power mains. Typical values from literature are $C_B \approx 300$ pF and $C_P \approx 20$ pF [Mars 84], [Pall 88]. These values serve as an example and can change considerably, for example, when touching a metal structure like a window frame.

As a result of the capacitive coupling, the body potential as a whole will oscillate with 50 or 60 Hz with respect to earth.

The voltage between the body and the system ground (the reference potential of the amplifier for bioelectric events) is called V_{CM} (common-mode voltage). The voltage between the system ground and earth is called V_{IM} (isolation mode voltage). The sum of both voltages is called V_{PO} (body voltage) and can easily attain values of several volts. If for example we consider the power-line voltage in Switzerland (230 V, 50 Hz) we can express the body voltage V_{PO} by:

$$V_{PO} \approx \frac{Z_B}{Z_P + Z_B} V_{230} \qquad (2.2)$$

with

$$Z_B = \frac{1}{j\omega C_B} \qquad (2.3)$$

$$Z_P = \frac{1}{j\omega C_P} \qquad (2.4)$$

follows

$$V_{PO} \approx \frac{C_P}{C_P + C_B} V_{230} \approx 14 \text{ V}_{\text{rms}} \qquad (2.5)$$

Z_B and Z_P denote the impedance associated with the capacitance versus earth and the power line respectively, j stands for the imaginary unit.

The amplifier for bioelectric events in Fig. 2.4 has two signal electrodes ($e1$ and $e2$) and one reference electrode (rl). Z_{e1}, Z_{e2} and Z_{rl} stand for the corresponding electrode-skin impedances. The input impedance of the amplifier is not yet shown. We will discuss the influence of the input impedance later (see 2.2.4).

The reference electrode (rl) connects the system ground to the body. Yet, this electrode is not mandatory; there are amplifiers for bioelectric events with only two electrodes (see 2.3.2)

The bioelectric signal of interest is the differential voltage between the two signal electrodes ($e1$ and $e2$), the corresponding voltage is denoted as V_{DM} (differential mode voltage).

The common-mode voltage V_{CM} is defined by the measurement situation:

$$V_{CM} = V_{PO} - V_{IM} \qquad (2.6)$$

The main source for the common-mode voltage V_{CM} is the power lines. If we consider again the power-line voltage in Switzerland (230 V, 50 Hz) we can express the common-mode voltage as:

$$V_{CM} = Z_{rl} \frac{Z_B Z_{sP} - Z_{sB} Z_P}{Z_{rl}(Z_P+Z_B)(Z_{sP}+Z_{sB})+Z_P Z_B(Z_{sP}+Z_{sB})+Z_{sP} Z_{sB}(Z_P+Z_B)} V_{230} \qquad (2.7)$$

Where $Z_B = 1/j\omega C_B$ stands for example for the impedance associated with capacitor C_B.

For surface electrodes made out of Ag/AgCl we can assume that at 50 Hz $Z_{rl} \ll Z_x$ with $Z_x \in \{Z_P, Z_B, Z_{sB}, Z_{sP}\}$ which leads to the simplified

2.2. Signal Acquisition

equation:

$$V_{CM} \approx Z_{rl} \frac{Z_B Z_{sP} - Z_{sB} Z_P}{Z_P Z_B(Z_{sP}+Z_{sB}) + Z_{sP} Z_{sB}(Z_P+Z_B)} V_{230} \qquad (2.8)$$

It is very difficult to give an approximate value for the resulting common-mode voltage V_{CM} because the values of Z_{sP} and Z_{sB} vary very much between different amplifiers and measurement situations. The value of V_{CM} can be anywhere in the range from 0 V to 230 V_{rms}.

It is important to note that according to equation (2.8) the amplitude of the common-mode voltage V_{CM} scales linearly with the electrode-skin impedance Z_{rl} of the reference electrode.

It is also informative to note that according to equation (2.7) there are two particular situations for which the common-mode voltage V_{CM} is zero.

$Z_{rl} = 0$: Unfortunately, this is not possible in a real measurement situation. For low frequencies there is always an electrode-skin impedance of several $k\Omega$. A good way to reduce the influence of Z_{rl} is to use a DRL (driven right leg) circuit as discussed in section 2.3.3.

$Z_B Z_{sP} = Z_{sB} Z_P$: If this equation is fulfilled then the numerator of equation (2.7) would be zero. Unfortunately, these capacities cannot be controlled. As a measure of precaution, the patient should avoid touching any metal surface during a recording because this will most probably lead to low values for either Z_B or Z_P and, most of the time, increase the inequality of the two terms in the equation above leading to even more interferences.

In all other circumstances there is a common-mode voltage $V_{CM} \neq 0$.

There is one more important case to consider:

$Z_{sB} \gg Z_B$; $Z_{sP} \gg Z_P$: This is a typical situation for body-worn portable devices which store the recordings locally or transmit the data via a wireless link. This setting will result in a relative low common-mode voltage.

For this setting the resulting formula reads:

$$\begin{aligned} V_{CM} &= Z_{rl} \frac{Z_B Z_{sP} - Z_{sB} Z_P}{Z_{sP} Z_{sB}(Z_B + Z_P)} V_{230} \\ &= j\omega Z_{rl} \frac{C_{sB} C_P - C_B C_{sB}}{C_P + C_B} V_{230} \approx 120 \; \mu V_{\text{rms}} \end{aligned} \qquad (2.9)$$

Typical values used were Z_{rl}=10 kΩ (@ 50 Hz), C_B=300 pF, C_P=20 pF, C_{sB}=300 fF and C_{sP}=200 fF. The power-line frequency was assumed to be f=50 Hz (Europe). Note, in this model the phase shift between V_{CM} and V_{230} is either +90° or −90°.

Again, the amplitude of the common-mode voltage V_{CM} scales linearly with the electrode-skin impedance Z_{rl} of the reference electrode.

As the bioelectrical signal V_{DM} is in the order of some millivolts or lower, it is important that the amplifier for bioelectric events effectively rejects the common-mode voltage. This feature is quantified by the CMRR (common mode rejection ratio), which is defined as the ratio between the differential-mode gain G_{DM} and the common-mode gain G_{CM} (see also section 6.3.3):

$$\text{CMRR} = \left|\frac{G_{DM}}{G_{CM}}\right| \qquad (2.10)$$

with

$$G_{DM} = \frac{\partial V_{out}}{\partial V_{DM}} \qquad (2.11)$$

$$G_{CM} = \frac{\partial V_{out}}{\partial V_{CM}} \qquad (2.12)$$

An amplifier for bioelectric events should have a CMRR of more than 80 dB (@ 50 Hz) [Mett 91]. This is necessary to suppress the common-mode voltage V_{CM} at the output of the amplifier to values lower than the smallest amplified signal that still is above the noise floor of the electrodes.

The second reason why the common-mode voltage has to be reduced is that, according to equation (2.2), the body voltage V_{PO} can reach values of about 14 V_{rms}. Yet, the supply voltage of the input stage of an amplifier for bioelectric events will be most probably below 14 $V_{rms} \approx 40\ V_{pp}$. Yet, the peak-to-peak value of the common-mode voltage V_{CM} must stay within the limits of the supply voltage, otherwise the input signal may be clipped. As a result, the common-mode voltage must be smaller than the 40 V_{pp} assumed for the body voltage V_{PO}. This is achieved by either a good isolation (e.g., a large Z_{sB} and Z_{sP}), a DRL circuit or both (for the DRL circuit please refer to section 2.3.3).

Body-worn devices which store the bioelectric recording locally, with or without transmission, are very well isolated, which results in a very low common-mode voltage V_{CM} as shown by equation (2.9). To reduce the common-mode voltage below the amplified input signal of some μV a CMRR of 50 dB is sufficient. A DRL circuit may reduce the common-mode voltage by 10 to 50 dB [Mett 90]. In practice, a reduction of 30 dB is easily achieved. Using a DRL circuit then reduces the CMRR requirement to 20 dB.

2.2. Signal Acquisition

Table 2.2 recapitulates the minimal CMRR (@ 50 Hz) we recommend for an amplifier for bioelectric events:

Tab. 2.2: Recommended CMRR for amplifier for bioelectric events. For the DRL circuit please refer to section 2.3.3

CMRR (@ 50 Hz)	DRL circuit will be used	without DRL circuit
mains powered	80 dB	—[a]
body-worn	20 dB	50 dB

[a] Not recommended because the amplifier may saturate due to excessive common mode interference

2.2.4 Common Mode to Differential Mode Conversion

The total CMRR of an amplifier for bioelectric events is, most of the time, limited by parasitic effects which are not under the full control of the designer. The four most important parasitic effects are described in the following sections, starting with the capacitive coupling to the leads.

Capacitive Coupling to the Leads

Fig. 2.5 depicts a typical measurement situation with two parasitic elements made visible: The parasitic capacitances C_{p1} and C_{p2} between the power line and the two leads as well as the input impedances Z_{i1} and Z_{i2} of the amplifier for bioelectric events.

For the discussion of the common-mode voltage we consider a measurement system which is grounded, i.e., $Z_{sB=0}$. This corresponds to a standard measurement system (e.g., an oscilloscope) and is a valid model to understand the different processes leading to a superposition of the common-mode voltage V_{CM} on the amplified bioelectrical signal V_{out}. The model cannot be used to explain the origin of the common-mode voltage (this was done in the previous section) and even more, such a system would not fulfill the patient safety requirements.

As a first result of $Z_{sB} = 0$ we can state that the system ground and earth have the same electric potential, i.e., $V_{IM} = 0$. If we further assume that $Z_{i1} \gg Z_{e1}$ and $Z_{i2} \gg Z_{e2}$ we can express the differential voltage $V_{DM\ para}$ as a result of the common-mode voltage V_{CM} and the two parasitic capacitances Z_{p1} and Z_{p2} by:

$$V_{DM\ para} = \left(\frac{Z_{e2}}{Z_{e2} + Z_{p2}} - \frac{Z_{e1}}{Z_{e1} + Z_{p1}} \right) (V_{230} - V_{CM}) \qquad (2.13)$$

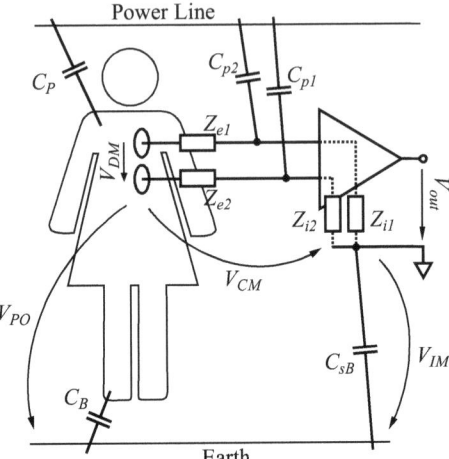

Fig. 2.5: Typical measurement situation where parts of the common-mode signal are first converted to a differential signal by the means of parasitic elements and then amplified by the system thus degrading the CMRR of the amplifier.

Where $Z_{p1} = 1/j\omega C_{p1}$ stands for example for the impedance of the parasitic capacitance C_{p1}.

The parasitic capacitances together with the corresponding electrode-skin impedances form two individual potential dividers. An asymmetry between these two dividers will result in part of the common-mode voltage V_{CM} being converted into a differential voltage $V_{DM\ para}$ and amplified by the system. This effect degrades the common mode rejection of the whole amplifier system without affecting the CMRR of the differential amplifier itself.

To describe this effect we calculate the CMRR resulting from the capacitive coupling which is expressed by:

$$\text{CMRR}_{para} = \left| \frac{\partial V_{CM}}{\partial V_{DM\ para}} \right| = \left| \frac{\partial V_{DM\ para}}{\partial V_{CM}} \right|^{-1} \quad (2.14)$$

$$= \left| \frac{(Z_{e1} + Z_{p1})(Z_{e2} + Z_{p2})}{Z_{e1}(Z_{e2} + Z_{p2}) - Z_{e2}(Z_{e1} + Z_{p1})} \right| \quad (2.15)$$

The resulting CMRR_{para} can achieve values from 20 dB to 200 dB (@ 50 Hz) depending on the topology of the amplifier. Values above 150 dB are not found in practice.

2.2. Signal Acquisition

According to equation (2.15) it is possible to reduce the influence of the parasitic capacitances C_{p1} and C_{p2} by reducing the electrode-skin impedance of the electrodes Z_{el} (e.g., by using active electrodes). Another possibility is to shield the wires with a low-impedance shield [Hors 98].

Note, shielding the wires with system ground will reduce the input impedance of the amplifier by increasing the input capacitance [Mett 91].

The potential-divider Effect

A very similar process is known as the *potential-divider effect* [Mett 90]. Again, there is a differential mode voltage $V_{DM\ pot}$ which is a result of the common-mode voltage V_{CM} and a parasitic effect, i.e., the imbalance of two electrode-skin impedances Z_{e1} and Z_{e2}. Referring again to Fig. 2.5 and neglecting the parasitic capacitances C_{p1} and C_{p2} we can write:

$$V_{DM\ pot} = \left(\frac{Z_{i2}}{Z_{i2} + Z_{e2}} - \frac{Z_{i1}}{Z_{i1} + Z_{e1}} \right) V_{CM} \qquad (2.16)$$

We will replace the individual impedances by a term using the mean value and the difference:

$$Z_{x1} = \overline{Z_x} + \frac{1}{2}\Delta Z_x \qquad Z_{x2} = \overline{Z_x} - \frac{1}{2}\Delta Z_x \qquad (2.17)$$

$\Delta Z_x = Z_{x1} - Z_{x2}$ stands for the difference between the two corresponding impedances and $\overline{Z_x}$ represents the mean value of the corresponding impedances with $x \in \{i, e\}$.

If we assume that $Z_{i1} \gg Z_{e1}$ and $Z_{i2} \gg Z_{e2}$ we can rewrite equation (2.16) as follows:

$$V_{DM\ pot} = \frac{\overline{Z_e}}{\overline{Z_i}} \left(\frac{\Delta Z_e}{\overline{Z_e}} - \frac{\Delta Z_i}{\overline{Z_i}} \right) V_{CM} \qquad (2.18)$$

The relative difference of the two input impedances $\Delta Z_i / \overline{Z_i}$ is in general much smaller compared to the relative difference of the electrode-skin impedances $\Delta Z_e / \overline{Z_e}$. We can therefore neglect the second term within the parenthesis of equation (2.18) which leads to the equation encountered throughout the literature:

$$V_{DM\ pot} \approx \frac{\Delta Z_e}{\overline{Z_i}} V_{CM} \qquad (2.19)$$

The resulting CMRR is expressed by:

$$\mathrm{CMRR}_{pot} = \frac{\overline{Z_i}}{|\Delta Z_e|} \qquad (2.20)$$

The relative variation between different electrode-skin impedances may easily reach 50 % and cannot be controlled. To minimize the potential-divider effect the input impedance of the amplifier for bioelectric events should be as high as possible.

Modern MOS-FET input stages, which at 50 Hz are mainly capacitive, achieve input impedances which are larger than 1 GΩ (i.e., $> 10^9 \Omega$).

For $\Delta Z_e = 10$ kΩ (@ 50 Hz) and $\overline{Z_i} > 10^9$ Ω (@ 50 Hz) the resulting upper limit for the CMRR will be over 100 dB (@ 50 Hz). At the stated frequency the input impedance $\overline{Z_i} > 10^9$ Ω corresponds to an input capacity $C_i < 3.2$ pF. Knowing that for a purely capacitive input the input impedance is given by

$$Z_i = \frac{1}{j\omega C_i} \qquad (2.21)$$

An alternative method often implemented in EEG recording systems is to actively measure the electrode-skin impedances at the beginning of the measurement and warn the operator if the impedance mismatch exceeds a certain limit.

Gain Mismatch in Amplifying Electrodes

This effect is usually only seen in amplifiers for bioelectric events using amplifying electrodes, e.g., active electrodes with a gain not equal to one, or digitizing electrodes. If the gain of two individual electrodes is different it results again a differential voltage $V_{DM\ gain}$ which is directly proportional to the common-mode voltage V_{CM}:

$$V_{DM\ gain} = \Delta G\, V_{CM} \qquad (2.22)$$

Where $\Delta G = G_1 - G_2$ stands for the difference of the two gains. The corresponding CMRR reads:

$$\text{CMRR}_{gain} = \left| \frac{G_{DM}}{G_{CM}} \right| = \left| \frac{\overline{G}}{\Delta G} \right| \qquad (2.23)$$

If two individual amplifying electrodes have a gain set by resistors with a certain tolerance Q this will limit the CMRR of a signal amplified by these two electrodes. For example, if we consider amplifying electrodes where the gain is set with two resistors R_1 and R_2 using a non-inverting configuration (see Fig. 6.1 for an example) we can assume that the most critical configuration concerning the CMRR is given by:

$$G_1 = 1 + \frac{R_1(1+Q)}{R_2(1-Q)} \quad \text{and} \quad G_2 = 1 + \frac{R_1(1-Q)}{R_2(1+Q)} \qquad (2.24)$$

2.2. Signal Acquisition

thus

$$\Delta G = \frac{R_1(1+Q)^2 - (R_1(1-Q)^2)}{R_2(1-Q^2)} \qquad (2.25)$$

with $Q^2 \ll 1$

$$\Delta G \approx \frac{4QR_1}{R_2} \qquad (2.26)$$

We can then calculate the worst-case value for the CMRR according to equation (2.23):

$$\text{CMRR}_{gain} = \frac{1}{4Q} \left| \frac{\overline{G}}{\overline{G}-1} \right| \qquad (2.27)$$

with

$$Q = \frac{|\Delta R|}{\overline{R}} \qquad (2.28)$$

First of all we see that for buffer electrodes ($G = 1$) there is no additional CMRR limitation due to the gain mismatch. If we now consider an active electrode with a gain of 20 dB (non-inverting configuration) we conclude that according to equation (2.27) using resistors with a tolerance of 1% would lead to a worst-case CMRR of merely 28.4 dB, far too low for an amplifier for bioelectric events. On the other hand, to guarantee a worst-case CMRR of 80 dB the tolerance of the employed resistors needs to be 0.025 ‰, far too high for individual components. To reach a good CMRR the gain mismatch must be compensated for (see 6.3.2).

Note, equation (2.27) does not apply to amplifying electrodes when the gain-setting network is connected to a common node (not the signal ground) corresponding to the mean value of all input voltages [Valc 04]. See also section 2.2.5 later in this chapter.

Corner-Frequency Mismatch for Individual Highpass Filters

Most amplifiers for bioelectric events employ a highpass filter. The filter removes the DC offset between different electrode-electrolyte interfaces which can attain several hundred millivolts (see 2.3.6). In many applications the DC offset is removed only after the bioelectric signal is converted into a single-ended signal. Thus both the signal and the highpass filter are referenced to the system ground.

If capacitive coupled electrodes are used (see section 2.2.1), or if the highpass is realized somewhere else in the circuit where the bioelectric signal is still bipolar, the tolerances of the elements forming the highpass filter will also lead to a limitation of the CMRR.

We can describe the transfer function of a highpass filter by the general form:

$$h(f) = \frac{jf\frac{1}{f_0}}{1+jf\frac{1}{f_0}} \qquad (2.29)$$

Where j is the imaginary unit and f_0 the corner frequency of the highpass filter. For bioelectric signals the CMRR is usually most limited at 50 Hz (or 60 Hz). For Switzerland we can therefore examine the gain of the highpass filter at 50 Hz and obtain:

$$G_{hp}(50\,\text{Hz}) = |h(50\,\text{Hz})| = \frac{50}{\sqrt{2500+f_0^2}} \qquad (2.30)$$

If individual highpass filters are used their gain at 50 Hz will be different depending on their actual corner frequency f_0 due to component tolerances. The corresponding CMRR limitation is calculated by equation (2.23).

Thus, the variation of the gain ΔG has to be estimated from equation (2.30):

$$\text{CMRR}_{hp}(@50Hz) = \frac{\overline{G}}{|\Delta G|} \approx \frac{G(f_0=\overline{f_0})}{|\Delta G|} \qquad (2.31)$$

according to Taylor's theorem (error propagation)

$$\Delta G \approx \frac{dG}{df_0}\Delta f_0 = -\frac{50 f_0}{\sqrt{(2500+f_0^2)^3}}\Delta f_0 \qquad (2.32)$$

$$\approx -G \frac{f_0^2}{2500+f_0^2}\frac{\Delta f_0}{f_0} \qquad (2.33)$$

it follows

$$\text{CMRR}_{hp}(@50Hz) \approx \frac{2500+\overline{f_0}^2}{\overline{f_0}^2}\frac{\overline{f_0}}{|\Delta f_0|} \qquad (2.34)$$

If, for example, the highpass filters are built using capacitors with a tolerance of 10% and resistors with a tolerance of 1% we can say that for the worst-case the corner frequency has a tolerance of 11%. For a corner frequency $f_0 = 1$ Hz (e.g., for EEG) we then calculate the worst-case value for the CMRR (@ 50 Hz) according to equation (2.34) and obtain 87.1 dB. For an ECG with a highpass filter at 0.016 Hz the same consideration will lead to a worst-case CMRR (@ 50 Hz) of 159 dB.

According to equation (2.34), the corner frequency has to be chosen as low as possible in order to minimize the effect of corner frequency variations. At the same time it can be estimated that the CMRR (@ 50 Hz)

2.2. Signal Acquisition

limitation due to component tolerances of individual highpass filters is not very critical.

It is important to note that this limitation also applies to electrodes with an individual *lowpass* filter like for example when using individual anti-aliasing filters for signal and reference electrodes or when using amplifiers with a limited bandwidth (see also chapter 4 and chapter 6). By analogy we can estimate that the corner frequency of individual lowpass filters should be at least 2.5 kHz (i.e., fifty times the power-line frequency) to again obtain a worst-case CMRR (@ 50 Hz) of 87.1 dB:

$$\mathrm{CMRR}_{lp}(@50Hz) \approx \frac{2500 + \overline{f_{0\,lp}}^2}{2500} \frac{\overline{f_{0\,lp}}}{|\Delta f_{0\,lp}|} \quad (2.35)$$

Where $f_{0\,lp}$ stands for the corner frequency of an individual lowpass filter.

2.2.5 Total Common Mode Rejection Ratio

As long as the bioelectric signal is differential, i.e., not converted into a single-ended signal, each stage in an amplifier for bioelectric events contributes with a finite CMRR. In the worst case the total CMRR of a sequence of stages in respect to a specific common-mode signal (e.g., the power-line interference) is given by [Pall 91b]:

$$\frac{1}{\mathrm{CMRR}_{total}} = \sum_i \frac{1}{\mathrm{CMRR}_i} \quad (2.36)$$

Where CMRR_i stands for the individual CMRR contribution due to one of the above mentioned effects. The total CMRR is always smaller than the smallest individual CMRR (in a worst-case scenario).

It is important to note that the total CMRR can also be better than the individual CMRRs (for a given frequency). This is the case if individual contributions of the same common-mode signal cancel each other (i.e., have a different sign). We will discuss a method to improve the CMRR of an amplifier for bioelectric events based on canceling contributions in detail in chapter 5

Most of the time one of the four described parasitic effects dominates. As a general rule of thumb we can estimate that for amplifiers which are not using a FET-input stage, the potential-divider effect will be limiting. For amplifying electrodes the gain mismatch will limit the CMRR. For capacitively coupled electrodes the corner-frequency mismatch is the dominant factor while for non-buffered EEG amplifiers the capacitive coupling to the leads will be most limiting.

CMRR of Instrumentation Amplifiers

This section is based on the publication of Pallas-Areny [Pall 91].

In section 2.2.4 we have seen that for amplifying electrodes the gain mismatch between individual electrodes can lead to a severe limitation of the CMRR. Also individual lowpass or highpass filter can lead to a limited CMRR (@ 50 Hz) when the corner frequency of the filter is close to the frequency of the power-line interference. The basic consideration of gain mismatch can also be applied to amplifiers. Fig. 2.6 shows three basic amplifiers, one DA (differential amplifier) and two versions of an INA (instrumentation amplifier).

The DA in Fig. 2.6 can also be seen as having two branches with a individual gain. Therefore, according to equation (2.23), a gain mismatch between these two branches will result in a limitation of the CMRR.

If the four resistors have a tolerance Q and the DA has a closed-loop gain of G the worst-case CMRR is then reduced to [Pall 91] :

$$\text{CMRR}_{gain} = \frac{G+1}{4Q} \qquad (2.37)$$

with

$$Q = \frac{|\Delta R|}{\overline{R}}$$

If for example resistors with a tolerance of 1 % are used for a differential amplifier with a nominal gain of 20 dB, the resulting CMRR_{gain} yields 48.8 dB. This is still much lower than the recommended value of 80 dB (@ 50 Hz) [Mett 93] but higher than the 28.4 dB corresponding to the worst-case CMRR for two individual amplifying electrodes with a gain of 20 dB (see section 2.2.4). This is because the gain of the differential amplifier improves the CMRR whereas the gain of the amplifying electrodes has very little effect on the CMRR.

Considering the INA depicted in Fig. 2.6 c) we can estimate the total worst-case CMRR based on resistor tolerances by combining equation (2.36), equation (2.37) and equation (2.27).

$$\frac{1}{\text{CMRR}_{total\ worstcase}} = \frac{1}{\text{CMRR}_{AE}} + \frac{1}{\text{CMRR}_{DA}}$$

$$= 4Q \frac{|\overline{G_{AE}} - 1|}{\overline{G_{AE}}} + \frac{4Q}{G_{DA}+1} \qquad (2.38)$$

with

$$Q = \frac{|\Delta R|}{\overline{R}}$$

2.2. Signal Acquisition

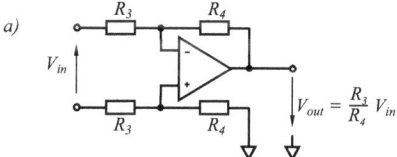

Differential Amplifier (DA)

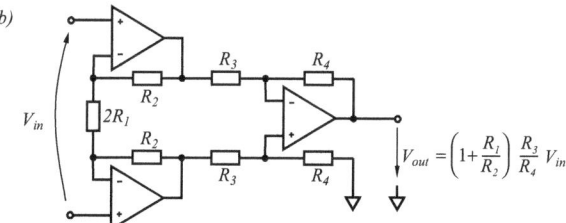

Instrumentation Amplifier (INA)
with coupled first stage

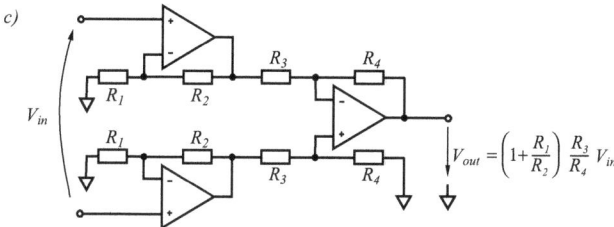

Instrumentation Amplifier (INA)
with non-coupled first stage

Fig. 2.6: Three basic topologies of amplifiers often used in bioelectric recordings. Inset a) shows a differential amplifier (DA) consisting of one op-amp and four resistors. The input resistance of the (DA) is given by two resistors in series. Inset b) and c) depict an instrumentation amplifier (INA) based on three op-amps. The INA offers a much higher input resistance than the DA. The first stage of the INA adds a gain to the DA which forms the second stage of the INA. The input stage in b) is coupled whereas the input stage in c) is not coupled. All three amplifiers convert a differential input signal into a single-ended output signal.

Where CMRR_{AE} stands for the CMRR of the amplifying electrodes (i.e., the first stage of the depicted INA with non-coupled input buffers) with a gain of G_{AE} and CMRR_{DA} for the CMRR of the differential amplifier (i.e., the second stage of the INA) featuring a gain G_{DA}. In a general case, both stages will superimpose part of the common-mode signal onto the output signal as a result of their finite CMRR. When these two interferences have the same sign they add up and the total CMRR will be dominated by the stage with the lower CMRR.

Note, the CMRR of the total amplifier can again be very high if the two interferences cancel each other due to an opposite sign. In chapter 5 and chapter 6 two circuits are presented who improve the total CMRR by controlling the CMRR of the second stage.

The INA with coupled input buffers depicted in Fig. 2.6 b) is different when it comes to the CMRR. Because the two input buffers are coupled it can be shown (see [Pall 91]) that the total CMRR is higher for the same resistor tolerance. Because of the coupling the effect of resistor tolerances can be neglected for the first stage. If the two input op-amps are matched, both in CMRR and open-loop gain A_0 the CMRR of the first stage is then very high and the total CMRR of the INA will be only limited by the second stage. The CMRR of the second stage on the other hand is improved by the gain of the first stage, hence, the worst-case CMRR for the INA with coupled input buffers from Fig. 2.6 b) would then be:

$$\mathrm{CMRR}_{total} \approx \left(1 + \frac{R_1}{R_2}\right)\left(1 + \frac{R_3}{R_4}\right)\frac{1}{4Q} \qquad (2.39)$$

with

$$Q = \frac{|\Delta R|}{R}$$

For example considering an amplifier for EEG where the gain of the first stage amounts to 20 dB, the gain of the second stage to 40 dB and the tolerance of the resistors used is 1% the worst-case CMRR due to resistor tolerances (i.e., gain mismatch) would be 88 dB. Note, this worst-case value does not depend on the signs of the two individual CMRR contributions.

2.2.6 Reducing the power-line Interference by Filtering

If the CMRR of an amplifier for bioelectric events is reduced by one or more of the above described parasitic effects, the power-line interference will appear superimposed on the bioelectric recording. As a result, important timing information may be impossible to read or clipping of the signal may occur due to amplifier saturation.

To reduce the power-line interference of the recording, a notch filter may be used to restore the original signal (as long there was no clipping of the signal). In our opinion, the use of a notch filter should be avoided whenever possible for two reasons:

- The notch filter does not differentiate between the power-line interference and the bioelectrical signal and thus removes part of the bioelectrical signal.

- An analog notch filter has a highly non-linear phase shift and thus alters the form of the bioelectric signal. This changes important timing information especially in ECG.

Instead of using a filter, a more elegant method is to generate an artificial sinusoidal voltage with the same phase, amplitude and frequency as the power-line interference and then subtract it from the bioelectric signal [Dots 96] and [Hwan 08]

A very good analysis which compares notch filters to adaptive filters is presented in [Hami 96].

It is always better to maximize the CMRR rather than filtering the signal afterwards. Maximizing the CMRR reduces the risk of clipping and removes all common mode interferences, the power-line interference itself but also its harmonics.

Note, we do not discourage filtering in general. An analog anti-aliasing filter should be used to remove the out-of-band component of the signal prior to sampling the bioelectrical signal. The sampled signal can then be filtered digitally to remove quantization noise from the sampling process. A smoothing filter which is well adapted to biological signals is the Savitzky-Golay filter which is available in MATLAB also. The Savitzky-Golay filter performs a local polynomial regression which has the advantage that local minima or maxima are better preserved than by a moving-average filter.

2.3 Amplifiers for Bioelectric Events

An amplifier for bioelectric events requires some means to define a reference for the bioelectric signal to measure. This reference is normally not the earth but a signal related to a body potential on the same body than the signal of interest. We now resume the most commonly used ways to generate a reference voltage for biopotential measurements.

2.3.1 Amplifiers using Tripolar Electrodes

If the biological signal of interest is localized, e.g., the electric activity of a particular muscle, a single electrode with three contacts may be used for the amplifier. Two contacts are used as an input to the system and serve to measure the signal of interest, one contact is used for the reference potential. An example for such a tripolar electrode is shown in Fig. 2.7 [Boni 95].

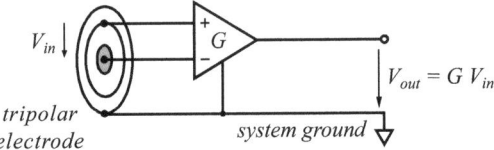

Fig. 2.7: Tripolar Electrodes use three distinct contacts on *one* electrode, two for the signal and one for the reference.

Where G stands for the differential gain of the amplifier.

Tripolar electrodes are used to measure well located differential signals as for example the electrical activity of a nerve at a particular spot [Rieg 03]. It is recommended designing the three contacts to have an equal surface area in order to have comparable electrode-skin impedances. Because of the small distance between the two signal contacts the measured voltage tends to be very small (some microvolts).

Tripolar electrodes are not suitable for surface EEG because in EEG all channels require the same reference potential to allow for a correct interpretation of the signal. This limitation also applies to multi-lead ECG.

2.3.2 Amplifiers for Two Electrodes

Amplifiers for bioelectric events which use only two signal electrodes and no reference electrode are very appealing due to the minimal number of contacts. A well-known example for such an amplifier is the ECG breast-belt from Polar. Integrated in this belt are an amplifier and two rubber electrodes for measuring the heart rate (but not the ECG).

In order for such an amplifier to work correctly, the two input potentials derived from the electrodes must lay within the two rails of the power supply. As we have discussed in section 2.2.3, the electric potential of the human body may be oscillating at the power-line frequency with an amplitude of several volts. An amplifier for bioelectric events must therefore have some means of controlling the common-mode voltage V_{CM} (see Fig. 2.4) between the body and the system ground of the amplifier.

This is for example possible by using a voltage divider which is placed between the two electrodes and connected to the system ground of the amplifier as shown in Fig. 2.8:

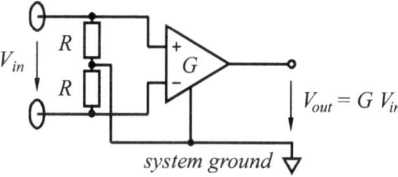

Fig. 2.8: An amplifier with two electrodes. The common-mode voltage is derived by a voltage divider.

The size of the two resistors R which build the voltage divider should be large in order to guarantee a large CMRR because of the potential-divider effect (see section 2.2.4). They also should be chosen larger than 1 GΩ in order to minimize the motion artifacts [Zipp 79b] (see 2.2.2).

Yet, the common-mode voltage is known to increase with the electrode-skin impedance of the reference electrode as seen in equation (2.8). Even if this equation is valid for a different topology we can expect that the underlying principle will also apply to amplifiers for two electrodes. And therefore we expect that the common-mode voltage V_{CM} will increase for large resistor values R of the voltage divider. Even more, due to the high impedance between body and system ground the common-mode voltage cannot be actively reduced.

As a result, most amplifiers for two electrodes show large power-line interferences. In [Spin 05] the optimal values for the input impedance of amplifiers for two electrodes are discussed in more detail.

Because of the relatively small input resistance R (when compared to FET-input amplifiers), amplifiers for two electrodes tend to have a low CMRR and suffer from larger motion artifacts. We, therefore, do not recommend the use of two electrode amplifiers for bioelectric events. The advantage of using only two electrodes instead of three electrodes does not compensate for an increased V_{CM} along with a reduced CMRR.

A recent publication describes a method for an amplifier with two electrodes to effectively reduce the common mode interference at the power-line frequency. The method employs a PLL (Phase-Locked-Loop) to generate a signal with the same amplitude but opposite phase which is then added to the amplified signal [Hwan 08].

The only industrial applications of two-electrode amplifiers known to us are defibrillators. Defibrillators require large electrodes with a very low electrode-skin impedance to avoid skin burns. In this particular application two-electrode amplifiers are used to detect the presence of the heart beat after a defibrillation pulse was administered. The low CMRR resulting from the two-electrode topology is of less concern as the electrodes are large, the electrode-skin impedance is small and no clinical ECG is to be measured.

2.3.3 Amplifiers using Reference Electrodes

We will now discuss systems which use one set of electrodes for the measurement of the bioelectric signal (called signal electrodes) and another set of electrodes for the common-mode voltage (reference electrode or/and DRL-electrode).

Amplifiers using an unbuffered Reference Electrode

The concept of the tripolar electrode (see section 2.3.1) can be generalized to an amplifier having two (or more) signal electrodes and one reference electrode. The latter can be used for example to directly connect the body surface potential to the signal reference of an amplifier for bioelectric events as shown in Fig. 2.9:

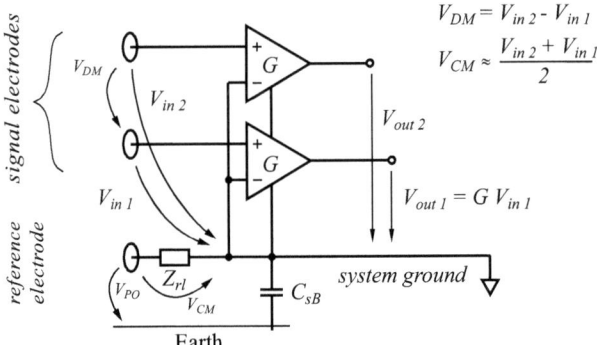

Fig. 2.9: An unbuffered electrode is placed unto the skin serving as reference electrode.

In Fig. 2.9 only two signal electrodes are shown, but the topology can easily be expanded to larger number of signal electrodes. In the case

2.3. Amplifiers for Bioelectric Events

of two signal electrodes the common-mode voltage V_{CM} as well as the bioelectric signal of interest V_{DM} can also be described by:

$$V_{DM} = V_{in\,2} - V_{in\,1} \qquad (2.40)$$

$$V_{CM} \approx \frac{V_{in\,2} + V_{in\,1}}{2} \qquad (2.41)$$

The value for the common-mode voltage V_{CM} given by equation (2.41) is an approximation which is only true if the bioelectric signals are neglected and thus the whole body would have the same potential. But it is a good approximation considering the small amplitude of bioelectric signals in general.

From Fig. 2.9 we can again see that the common-mode voltage V_{CM} would be zero if the electrode-skin impedance Z_{rl} of the reference electrode is zero. Unfortunately, this is never the case for an amplifier for bioelectric events because of the low conductivity of the skin.

If we take as an example an amplifier for bioelectric events using an isolation amplifier with the supply being connected to earth we can assume a relatively large capacitive coupling from the system ground to earth. If we take for example the ISO124 (orig. Burr-Brown, now TI) the isolation capacitance C_{sB} is about 2 pF. This seams a small value but in the case of multi-electrode amplifier this value quickly increases. If for example we take an EEG amplifier with 64 electrodes the isolation capacitance could easily attain 128 pF.

According to equation (2.8) we can estimate the common-mode voltage for the example above to:

$$V_{CM} \approx j\omega Z_{rl} \frac{C_P C_{sB} - C_B C_{sP}}{C_{sB} + C_{SP} + C_P + C_B} V_{230} = 4 \text{ mV}_{\text{rms}} \qquad (2.42)$$

Typical values used were Z_{rl}=10 kΩ (@ 50 Hz), C_B=300 pF, C_P=20 pF, C_{sP}=200 fF and a power-line frequency of 50 Hz.

Compared to the EEG signal itself which can have an amplitude of about 10 μV only, this is a large common-mode signal. Especially for mobile applications where the electrode-skin impedance can be larger than in the example above. To reduce the common-mode voltage actively one possibility is to omit the reference electrode and use a resistive network to generate a reference potential out of all input electrodes. This reference is then fed back to the body using a low impedance electrode. This is commonly done using a DRL circuit.

Amplifiers using a DRL Electrode

To reduce the impact of the electrode-skin impedance Z_{rl} of the reference electrode in Fig. 2.9 an active circuit may be used in order to drive the body surface potential. An example of such a system is the DRL (driven right leg) circuit[1]. The purpose of the DRL circuit is to reduce the common-mode voltage via negative feedback [Wint 83b]. An example of a DRL circuit is shown in Fig. 2.10.

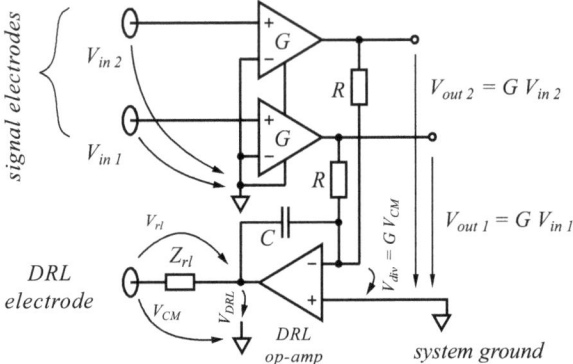

Fig. 2.10: A feedback loop is built around a driven right leg (DRL) electrode in order to force the system ground to a known potential, e.g., the potential of the body.

In Fig. 2.10 the amplified input voltages of the two signal electrodes are averaged using a resistive voltage divider built by the two resistors R. We will calculated the voltage using the *system ground* as the reference and obtain:

$$V_{div} = (V_{out\,1} + V_{out\,2})/2$$
$$= G\,(V_{in\,1} + V_{in\,2})/2 = G\,V_{CM} \qquad (2.43)$$

The voltage V_{div} generated by the voltage divider corresponds to the amplified common-mode voltage $G\,V_{CM}$ and is compared by the DRL op-amp to the system ground. The difference is amplified, inverted, integrated and fed back to the body via the DRL electrode. Thus, the voltage at the output of the DRL op-amp can then be expressed by:

$$V_{DRL} = -G\,\frac{2}{j\omega C R}\,V_{CM} \qquad (2.44)$$

It is important to note that equation (2.44) is only true as long as the output voltage of the DRL op-amp does not saturate and as long all signal electrodes are connected (to sense the common-mode voltage). This seems difficult at first because of the high gain of the DRL loop. But at

---

[1] Originally the DRL circuit was attached to an electrode placed on the right leg, thus the name of the circuit.

2.3. Amplifiers for Bioelectric Events

the same time the voltage at the input of the DRL op-amp (corresponding to the amplified common-mode voltage) will be driven to very low values by the negative feedback due to the principle of virtual ground.

The amplifier with the DRL circuit senses, amplifies, inverts and integrates the common-mode voltage V_{CM} and then presents the resulting voltage at the opposite pole of the electrode-skin impedance of the DRL electrode (acting as reference electrode). The voltage V_{rl} over the skin-electrode impedance Z_{rl} can be calculated as:

$$V_{rl} = V_{CM} - V_{DRL}$$
$$= \left(1 + G\frac{2}{j\omega CR}\right) V_{CM} \quad (2.45)$$

As a result of the DRL loop, the voltage over the impedance Z_{rl} is amplified which means the current through the impedance Z_{rl} is larger for the same common-mode voltage in the presence of a DRL circuit when compared to the previous circuit without DRL circuit (see previous section 2.3.3).

The larger current through the electrode-skin impedance with the same common-mode voltage can also be interpreted as if the electrode-skin impedance Z_{rl} would be smaller in the case of a DRL circuit. According to equation (2.7) the common-mode voltage V_{CM} scales linearly with Z_{rl}. Thus, by analogy we can conclude that the DRL circuit reduces the common-mode voltage by the factor by which the current through the electrodes-skin impedance increases for any given common-mode voltage V_{CM} (as long the DRL amplifier does not saturate).

The reduction of the common-mode voltage V_{CM} is often expressed as an improvement of the CMRR. The improvement of the CMRR due to a DRL circuit is then expressed by:

$$\text{CMRR}_{DRL} = \text{CMRR}_{orig} + 20\log\left(1 + G\frac{2}{j\omega CR}\right) \quad (2.46)$$

Where CMRR_{orig} is the CMRR without DRL circuit, G is the forward gain of the signal electrodes and $2/j\omega CR$ corresponds to the gain of the DRL circuit.

The reduction of the common-mode voltage V_{CM} is proportional to the gain of the feedback loop. A high gain of the feedback loop is therefore most desirable but may lead to instability. The role of the capacitor C in the feedback loop is to ensure stability by increasing the phase margin. To visualize the gain of the DRL circuit and to evaluate the phase margin we recommend the use of a P-Spice-based tool for every new circuit. An example of a CMRR measurement of an amplifier including a DRL circuit will be discussed later in section 6.3.3.

When using a DRL circuit, the common-mode voltage V_{CM} resulting from the power-line interference can be reduced by a factor of up to 300, resulting in an increase of the CMRR by about 50 dB (@ 50 Hz) [Mett 90].

For the circuit above there is no limitation to the number of signal electrodes. However, if one electrode does not connect to the body, this electrode acts like an antenna picking up power-line interference. This interference is then amplified and added to the amplified common-mode voltage via the resistive divider. For this reason systems with a large number of signal electrodes may preferably use one or more dedicated reference electrodes for the measurement of the common-mode voltage.

Using dedicated Electrodes for the DRL Loop

Instead of generating a reference by averaging all input signals it is possible to use one or more dedicated reference electrodes. Fig. 2.11 depicts a system with one dedicated reference electrode and two signal electrodes. The reference electrode senses the common-mode voltage V_{CM} As before, the common-mode voltage is amplified, inverted, integrated and fed back to the body in order to drive the common-mode voltage V_{CM} to lower values.

In Fig. 2.11 the reference electrode is buffered in order to generate a low-impedance node which drives the resistor R. It also has the effect that all electrodes present the same input impedance thus reducing the 'potential-divider effect' described in section 2.2.4.

If the reference electrode is affected by a bioelectrical signal, this signal appears superimposed on all channels. Thus, a reference electrode needs to be attached to a location which is ideally not affected by any bioelectric signal. It is also possible to use several electrodes connected by a resistive network. This reduces the sensitivity to interferences of the reference electrodes but again increases the risk of one reference electrode loosing contact.

Other uses of the DRL Circuit

The primary use of the DRL circuit is to reduce the common-mode voltage V_{CM}. The DRL circuit can also be used to force the common-mode voltage V_{CM} to a known signal by superimposing a voltage source to the reference of the DRL op-amp (instead of signal ground). The source can be a DC voltage [Levk 82] or an AC voltage [Ober 82].

Two examples for amplifiers using the DRL circuit for these purposes will be discussed later in this thesis. In chapter 3 we will use a DRL

2.3. Amplifiers for Bioelectric Events

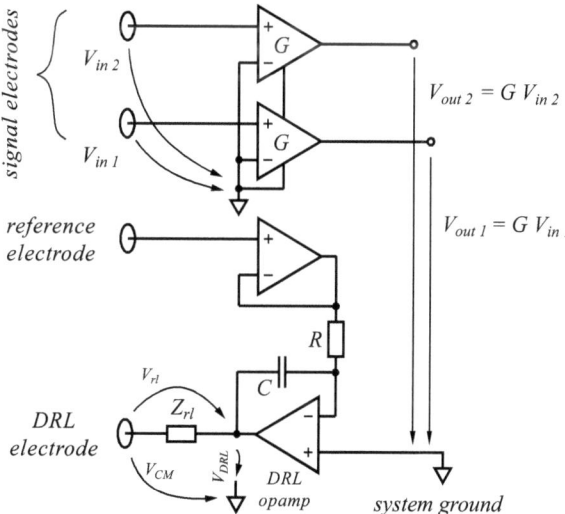

Fig. 2.11: A dedicated reference electrode is used to obtain a reference voltage.

circuit to superimpose a sinusoidal voltage of 10 kHz to measure the electrode-skin impedance mismatch. In chapter 6 we will demonstrate how a DRL circuit can be used to drive the DC value of the input voltage to the mid-range of an active input circuit, e.g., an active electrode.

2.3.4 AC-coupled Amplifiers

Most amplifiers for bioelectric events are designed with a highpass filter at some place in the signal path in order to remove the unwanted electrode-electrolyte potential, which may differ between electrodes by up to ±300 mV.

One possibility is to place the highpass filter after the differentiation, where the bioelectric signal is already converted from a bipolar signal to an unipolar signal. The disadvantage of this approach is that only a limited gain may be implemented in the stages prior to the highpass filter or else clipping of the signal may occur.

To avoid this limitation, some amplifiers place the highpass filter before the differentiation where the signal is still bipolar. These systems are generally called AC-coupled amplifiers. AC-coupled input stages are described since the early beginning of ECG recordings [Pall 89]. The topology is appealing because the maximum gain can be realized in the first stage. This minimizes the noise contribution of all following stages.

Examples for AC-coupled amplifiers are systems with a blocking capacitor right on the electrodes or systems which use insulating electrodes (see 2.2.1).

Fig. 2.12 shows a model of an insulated electrode where a coupling capacitor is placed in series with the signal path.

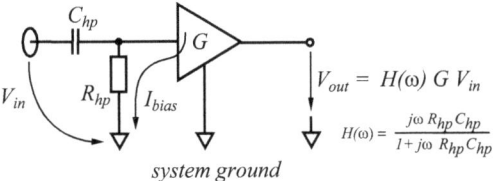

Fig. 2.12: Schematic view of an AC-coupled input stage

The corner frequency of this particular input stage is given by:

$$f_0 = \frac{1}{2\pi C_{hp} R_{hp}} \qquad (2.47)$$

The resistor R_{hp} is required to generate a path to the system ground for the input bias current I_{bias} of the amplifier. The potential-divider effect (see section 2.2.4) has the potential to limit the CMRR of AC-coupled amplifiers because the capacitor C_{hp} increases the source impedance of the signal while the resistor R_{hp} reduces the input impedance of the amplifier (here an op-amp with the gain G). The use of a bootstrap circuit can raise this input impedance at 50 Hz [Pall 90].

Many authors claim that using capacitively coupled electrodes reduces motion artifacts [Aliz 96], [Tahe 94]. We, however, have not been able to reproduce any measurements confirming a reduction of motion artifacts when using capacitively coupled electrodes.

Remote Off-Body ECG Sensing

Off-body measurement systems are a subgroup of AC-coupled amplifiers. As the name suggests, in off-body sensing systems the electrodes are *not* in contact with the body.

The first amplifier using off-body electrodes was described in 1993 by Clippingdale [*Clip* 93]. The development was driven by the need to

2.3. Amplifiers for Bioelectric Events

record ECGs from people with skin burns who could not tolerate electrodes in contact with the skin. Consequently, the electrodes for the ECG were placed under a hospital bed. The coupling between body and electrodes was purely capacitive, with the electrode forming one conductor of the capacitor and the human body forming the second conductor[1].

Because the capacitive coupling between electrode and skin is achieved through air at a distance of several centimeters, the resulting capacitance tends to be rather small, approximately given by the formula for a parallel-plate capacitor ($\sqrt{A} \gg d$ for a circular plate):

$$C \approx \frac{\varepsilon \varepsilon_0 A}{d} \qquad (2.48)$$

Where ε_0 is the electrical permittivity of free space, ε the electrical permittivity of the dielectric (if any), A the surface and d the distance in between of the two conductors.

Because of the small input capacitance the input resistance of the amplifier for bioelectric events has to be as large as possible in order to achieve a low corner frequency in the order of some Hertz. First, a bootstrap circuit was used to guarantee a large input impedance. Later, modern instrumentation amplifiers were used with virtually no bias current [Harl 02]. The input node of the amplifier is floating yet still constant over a long period of time [Harl 02b]. To avoid saturation of the floating input of the instrumentation amplifier the electrodes require elaborate shielding.

We would expect that the measurement is free of any motion artifact because the electrodes are no longer in direct contact with the skin. Unfortunately this is not the case, a motion of the body will also vary the size of the coupling capacity. Varying the size of a charged capacitor will generate a voltage change over the capacitor. To understand this, we assume the charge Q to be constant, which is true for a very short period of time. The voltage V over the capacitor is given by the charge divided by the capacitance.

$$V = \frac{Q}{C} = \frac{Q}{\varepsilon \varepsilon_0 A} d \qquad (2.49)$$

When the distance d between the skin and the electrode varies due to motion a voltage change results:

$$\Delta V = \frac{Q}{\varepsilon \varepsilon_0 A} \Delta d \qquad (2.50)$$

This voltage change is directly superimposed onto the bioelectric signal and hence not distinguishable from it. In [Kim 04] it was for example

---

[1] A capacitor can be described as being a passive electronic component consisting of a pair of conductors separated by an insulator

shown that the measured off-body recording showed waves which were due to the movement of the blood vessels and not due to the bioelectric activity of the heart. Note, although the signal is not of bio*electric* origin, it can still be used to measure the heart rate using an amplifier for bioelectric events.

To reduce the effect of body motion, a larger distance d is advantageous. This is because the relative voltage change is equal to a relative change in distance. From equation (2.50) we can deduce:

$$\frac{\Delta V}{V} = \frac{\Delta d}{d} \qquad (2.51)$$

Yet, a larger distance d yields a lower capacity and consecutively requires a larger input resistance.

In addition, off-body measurements suffer from a low CMRR because the difference between the different coupling capacitances can be very large, depending on the actual skin-electrode distance. The CMRR (@ 50 Hz) is then expressed by equation (2.34). In addition, the coupling capacitance correspond the skin-electrode impedance seen from the point of view of the potential-divider effect and thus equation (2.20) does also apply. As a consequence we highly recommend the use of a DRL circuit also for off-body amplifiers as demonstrated in [Lim 06], [Burk 00].

Because remote off-body systems also measure the movement of the surface of the skin they are not suitable for clinical ECG or EEG measurements. However, they offer very promising solutions for the observation of the heart beat (and therefore heart beat variation) with the greatest comfort possible, even in the bathtub [Lim 04] and without touching the skin.

Noise of AC-coupled Input Stages

Amplifiers for bioelectric events used in high-quality recordings should have an input-referred spectral voltage-noise density which is lower than the measured spectral voltage-noise density of an Ag/AgCl electrode (see Fig. 2.2). Contrariwise, amplifiers for mobile applications may have a higher input noise to allow for a reduced power consumption.

To investigate the noise behavior of an AC-coupled input stage we use a very simple model consisting of an ideal source V_{in} (source resistance is zero), a highpass filter and an ideal amplifier (no input noise and infinite input resistance). For bioelectric amplifiers the source resistance (i.e., the electrode-skin impedance) can be neglected because the input impedance of the amplifier is several magnitudes larger over the

2.3. Amplifiers for Bioelectric Events

full bandwidth of interest. The corresponding schematic is depicted in Fig. 2.13.

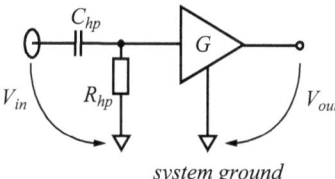

Fig. 2.13: Schematic view of a very simple model of an AC-coupled input stage

The capacitor C_{hp} of the highpass filter will shape the thermal voltage noise of the corresponding resistor R_{hp}. We can describe the spectral voltage-noise density at the input of the amplifier by:

$$\bar{v}_n = \begin{cases} \sqrt{4k_BTR_{hp}} & \text{if } 2\pi fC_{hp}R_{hp} \ll 1 \\ \frac{1}{2\pi fC_{hp}}\sqrt{\frac{4k_BT}{R_{hp}}} & \text{if } 2\pi fC_{hp}R_{hp} \gg 1 \end{cases} \quad (2.52)$$

Note, the above calculated spectral voltage-noise density has its origin in the resistors only, in theory capacitors are considered noiseless. This spectral voltage-noise density should stay below the spectral voltage-noise density of an Ag/AgCl electrode.

In addition, amplifiers for bioelectric events should have an input resistance of over 1 GΩ. In our model we then require that the resistor R_{hp} of the highpass filter would also be above several GΩ as the resistor will be in parallel to the input impedance of the amplifier and should not decrease the input resistance.

To verify that the spectral voltage-noise density of the highpass filter remains below the spectral voltage-noise density of the Ag/AgCl electrode, we therefore have to consider only the second part of equation (2.52) for which the spectral voltage-noise density decreases with $1/f$ similar to the spectral voltage-noise density of an Ag/AgCl electrode for frequencies below 3 Hz (see Fig. 2.2).

The asymptote of the electrode noise (Ag/AgCl) below 3 Hz drawn in Fig. 2.2 can now be compared to the second part of equation (2.52) for example at the frequency $f = 1$ Hz. From the graph we can then estimate:

$$\frac{1}{1\text{Hz} \cdot 2\pi C_{hp}}\sqrt{\frac{4k_BT}{R_{hp}}} \leq 70 \text{ nV}/\sqrt{\text{Hz}} \quad (2.53)$$

Solved for $C_{hp}\sqrt{R_{hp}}$ this yields:

$$C_{hp}\sqrt{R_{hp}} \geq 2.93 \cdot 10^{-4}\, \Omega^{-1/2}\, \text{Hz}^{-1} \qquad (2.54)$$

Multiplying both sides with $\sqrt{R_{hp}}$ results in:

$$C_{hp}R_{hp} \geq 2.93 \cdot 10^{-4}\, \Omega^{-1/2}\, \text{Hz}^{-1}\sqrt{R_{hp}} \qquad (2.55)$$

We can express the product on the left side by the corner frequency f_0 of the highpass filter which is defined by the application (e.g., ECG or EEG) by $C_{hp}R_{hp} = 1/2\pi f_0$. After substituting the product we can calculate the maximal resistance R_{hp} which still fulfills equation (2.55):

$$R_{hp} \leq \left(\frac{1}{2\pi f_0\, 2.93 \cdot 10^{-4}}\right)^2 \Omega \qquad (2.56)$$

For $f_0 = 0.016$ Hz (i.e., for ECG) we obtain $R_{hp} \leq 1.16$ GΩ and a corresponding $C_{hp} \geq 8.6$ nF.

This is a reasonable value for the input impedance of an amplifier for bioelectric events when used for ECG. For example the INA121 (TI) has an input stage which can be modeled by an input resistance $R_{in} = 1$GΩ parallel to an input capacitance of $C_{in} = 1$ pF. For frequencies below the corner frequency $f_0 = 0.016$ Hz the input impedance still behaves like a resistance and the highpass filter made out of R_{hp} and C_{hp} will still be functional. For higher frequencies (e.g., 50 Hz) the two capacitors C_{hp} and C_{in} will form a capacitive voltage divider with the gain of $C_{hp}/(C_{hp} + C_{in}) \approx 1$ which leads to a slightly reduced input signal.

But for $f_0 = 1$ Hz (EEG) we obtain $R_{hp} \leq 296$ kΩ which is far too low for an amplifier for bioelectric events. Yet, increasing R_{hp} will either increase the spectral voltage-noise density (for the same f_0) or result in a lower f_0 (for the same spectral voltage-noise density).

We conclude that AC-coupled amplifiers may be used for ECG measurements but are not suited for EEG measurements because there will always be a trade-off between spectral voltage-noise density and the input resistance or settling time.

As an example, in [Pran 00] and [Harl 02] an AC-coupled amplifier is built using the input resistance R_{in} of an INA 116 (orig. Burr-Brown, now TI) as part of the highpass filter. Noise measurements are provided as well.

2.3.5 DC-coupled Amplifiers

After all, there is a crucial downside to using AC-coupled input stages: The low-frequency part including the DC voltage of the bioelectric signal is inevitably lost. Yet, it was shown that removing the DC offset

2.3. Amplifiers for Bioelectric Events

may also remove some clinically relevant data. Examples of clinically relevant low-frequency signals are the depolarization prior to an epileptic seizure [Iked 99], the change of alertness that can be measured by EEG [Zsch 02] or the sleep pattern of preterm infants [Vanh 02].

Fig. 2.14 illustrates the effect of a highpass filtering by showing the same recording, once with a highpass filter and once without.

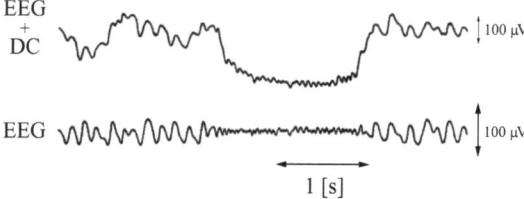

Fig. 2.14: Twice the same EEG recording, the first trace without a highpass filter and the second trace with a highpass filter. (*Picture reproduced from [Zsch 02]*)

Without discussing the medical meaning we can easily see that the trace without AC-coupling (named 'EEG + DC') holds more details than the trace with AC-coupling (named 'EEG'). The interpretation of the additional information in DC-coupled EEG systems is still under investigation. Yet, the fact that there is additional information available is undisputed.

This leads to a new emerging standard, the Full-Band EEG [Vanh 05]. To record a Full-Band EEG the amplifier must be built without a highpass filter. However, it is possible to allow for manual compensation of a DC offset at any given time by the operator.

Instead of removing the DC offset manually, it is possible to sample the full signal with an ADC having a large dynamic range. The required number of bits is given by the maximum tolerable offset divided by the desired resolution. If, for example, the maximum tolerable offset is ±300 mV as proposed by the AAMI [AAMI 99] and the desired resolution is 1 μV_{rms}, we can calculate the required number of bits:

$$n \geq \log_2 \left(\frac{600 \ mV}{1 \ \mu V} \right) = 19.2 \qquad (2.57)$$

An ADC with 20-bit resolution would be sufficient. This is for example the effective number of bits of the AD7734, a 24-bit ADC from Analog Devices, when used with a sampling rate of about 1 kHz.

We will demonstrate in chapter 6 a second way to measure the low-frequency part of a bioelectric signal, i.e., by looking at the feedback signal used to compensate for the skin-electrode offset.

Noise of DC-coupled Input Stages

Amplifier for bioelectric events require a very high input impedance (@ 50 Hz) to avoid limitation of the CMRR due to the potential-divider effect (see 2.2.4). They also require a very high input resistance to limit motion artifacts [Zipp 79].

For this reason most of the amplifiers for bioelectric events have FET input stages which are affected by $1/f$ noise. This means that the spectral voltage-noise density of the input-referred voltage noise is flat down to a certain corner frequency. Below this corner frequency the spectral voltage-noise density increases by a factor of ten per two decades. (The name $1/f$ is derived from the fact that the corresponding power spectral density rolls off with $1/f$).

Because an Ag/AgCl electrode has a similar characteristic (see Fig. 2.2), the amplifier can be designed to stay below the spectral voltage-noise density of an Ag/AgCl electrode.

2.3.6 Review of Amplifiers for Bioelectric Events

Reviewing the discussed amplifiers for bioelectric events we distinguish between four main topologies as depicted in Fig. 2.15.

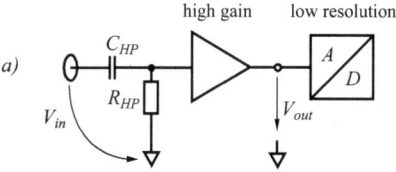

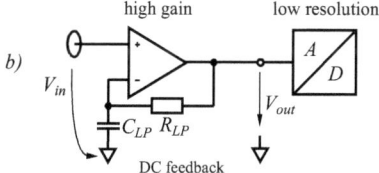

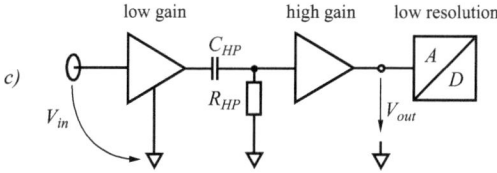

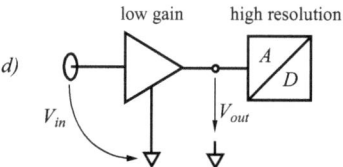

Fig. 2.15: Four main topologies for amplifier for bioelectric events. They are a) AC-coupled input stage, b) DC-coupled input buffer with DC feedback, c) standard two-stage bioelectric amplifier and d) DC-coupled input stage followed by a high-resolution ADC.

All four topologies are described in literature and used in commercially available amplifiers for bioelectric events. None of the four topologies is ideally suited for all purposes. The four topologies are:

a) AC-coupled input stage: This topology may suffer from an additional CMRR limitation due to the corner-frequency mismatch (see 2.2.4). Input impedance, input-referred noise or settling time may be a limiting factor especially for EEG (see 2.3.4). This is the only topology which allows off-body sensing of an ECG (see 2.3.4)

b) DC-coupled input buffer with DC feedback: In contrast to the AC-coupled input buffer the DC feedback topology allows to use a common reference electrode. The highpass filter is realized in the first stage which allows for a high gain in the first stage to reduce the noise contribution of the following stages. The DC gain is always one, thus the low-frequency part of the signal is not fully removed. As a result, this part of the signal can also be sampled in order to measure the bioelectric signal down to DC. This is the best topology for amplifying electrodes (see chapter 6) and [Mett 94]. This topology can also be used to build fully differential amplifiers which do not suffer from the CMRR limitation due to corner-frequency mismatch [Spin 03].

c) Standard two-stage amplifier: This presents the most frequently used topology in amplifiers for bioelectric events. A low-gain DC-coupled buffer is followed by a highpass filter and a second gain stage. The second gain stage boosts the bioelectric signal and allows the use of standard low-resolution ADCs or plotters. The gain of the first stage is limited to a factor of 5 to 15 (depending on the supply voltage) in order to handle DC offsets of ± 300 mV [AAMI 99] without saturating the first stage. The DC offset is normally rejected without being measured.

d) DC-coupled input stage followed by a high-resolution ADC: After an initial gain of 5 to 15 the bioelectric signal is sampled by a high-resolution ADC. This topology replaces more and more of the two-stage amplifiers because of the lower part count and the presence of the low-frequency part of the bioelectric signal in the measurement. This is only possible due to the development of ADCs with 24-bit resolution. Because most of the signal processing is performed digitally it offers the highest flexibility but also the highest power consumption of all systems presented. Filters are usually implemented as FIR (finite impulse response) filters with a perfect linear phase response resulting in a minimal distortion of the bioelectric signal. To our knowledge, there are no publications describing such a system in detail, yet several systems are available commercially on the market, e.g., 'active-one' from biosemi [bios].

2.4 Figures of Merit

At this point we would like to summarize the most important figures of merit for amplifiers for bioelectric events like EEG and EEG:

Noise: A high-quality amplifier for bioelectric events should generate less equivalent input noise than the employed electrode. According to Fig. 2.2 the spectral voltage-noise density of an Ag/AgCl-Electrode can be estimated to 30 nV/$\sqrt{\text{Hz}}$ at 100 Hz. The total noise power in a bandwidth of 10 Hz - 1 kHz (ECG) can be estimated as 1.4 μV_{rms} and the spectral voltage-noise density is inversely proportional to f for frequencies below 3 Hz.

Input-stage design: For a correct bioelectric signal representation the AAMI [*AAMI* 99] defines a lower corner frequency of at most 0.016 Hz for clinical ECG and 0.05 Hz for ambulant ECG. The corner frequency for EEG should be at most 1 Hz. The AAMI also defines a minimum level of ± 300 mV DC offset which the amplifier for bioelectric events must be able to accept. The offset originates mostly from the electrode-skin interface. To reduce motion artifacts, the input resistance of the amplifier should be larger than 1 GΩ [Zipp 79]. All filters should have a linear phase response to guarantee optimal signal reproduction.

CMRR: The CMRR (@ 50 Hz) has to be large enough to ideally reduce the power-line interference to a level below the resolution of the amplifier while in standard mode of operation (e.g., the person is not touching a metal surface). The required value for the CMRR depends on the measurement situation. A body-worn device with batteries and wireless link requires a CMRR (@ 50 Hz) of only about 20 dB (resp. 50 dB if no DRL is used) because there is only a very small common-mode voltage. A standard EEG device which is not entirely worn on the body should have a CMRR (@ 50 Hz) of at least 80 dB [*Mett* 93] and use a DRL circuit.

Soft factors: Mobile devices intended for the consumer market require the lowest possible cost for the components. This works hand in hand with the aim to achieve the lowest possible power consumption, because the autonomy increases and/or the weight of the device can be reduced by using smaller batteries.
Additional cost factors include the complexity of the electronic circuit (number of ICs) and required mechanical parts (number of wires and connectors).

3. CONTINUOUS MONITORING OF ELECTRODE-SKIN IMPEDANCE MISMATCH DURING BIOELECTRIC RECORDINGS

This chapter is a synopsis of the corresponding publication in the IEEE Transactions on Biomedical Engineering [Dege 08].

3.1 Problem Statement

The electrode-skin impedance is an important parameter for amplifiers for bioelectric events. A large electrode-skin impedance results in more interferences (noise) like for example increased coupling of the power-line voltage to the leads. A large electrode-skin impedance *mismatch* leads to a reduced CMRR.

It would be beneficial to monitor the electrode-skin impedance during the measurement in order to give feedback to the operator. The monitoring of the electrode-skin impedance should not degrade the quality of the recording. The following work describes a suitable method to continuously monitor the electrode-skin impedance.

3.2 Prior Art

Several publications describe how to measure electrode-skin impedances by injecting a sinusoidal current via an additional reference electrode through the body and then measuring the voltage drop over all the different signal electrodes. The frequency of the sinusoidal current ranges from 120 Hz [Hami 00] to 20 kHz [Devl 84]. These methods do not measure one electrode-skin impedance but two in series (the electrode-skin impedance of the reference electrode is always measured along with the electrode-skin impedance of the signal electrode under investigation).

Another method measures the electrode-skin imbalance based on the power-line interference [Spin 06] while the reference electrode has to be disconnected.

Grimbergen et al. describe a method to measure a single electrode-skin impedance based on the power-line interference while concurrently measuring the input current and the input voltage by a second high-impedance electrode [Grim 92].

Only one of these methods is designed to simultaneously measure the bioelectric signal [Devl 84]. Yet, all these methods lower the input resistance of the amplifier for bioelectric events, which degrades the measurement due to the potential-divider effect (see section 2.2.4) and increases motion artifacts [Zipp 79b].

Only one publication describes a method which does not reduce the input resistance of the amplifier for bioelectric events; it forces a known common-mode voltage V_{CM} via the DRL circuit [Ober 82]. Unfortunately, the method only yields a binary decision for 'bad' or 'good' contact. Our method is derived from this work, but is able to measure the electrode-skin imbalance between two electrodes.

3.3 Method

Our method is based on superimposing a known voltage V_{add} to the reference voltage of the DRL circuit (see also section 2.3.3) to force the common-mode voltage V_{CM} to a known voltage. The superposition of an additional signal does not affect the ability of the DRL to reduce the power-line interference.

The amplifier is considered to be a CMOS amplifier with a FET-input stage. This is the standard for amplifiers for bioelectric events. They have a very high input impedance Z_{in} modeled by a resistor R_{in} in parallel to a capacitor C_{in}:

$$Z_{in} = \frac{R_{in}}{1 + j\omega R_{in} C_{in}} \quad (3.1)$$

Typical values are $R_{in} > 10^{12}$ Ω and $C_{in} = 1$ pF.

The amplitude and frequency of V_{add} is not defined by the method but by the application and the bandwidth of the amplifier for bioelectric events. In other words, the frequency of V_{add} has to be chosen within the signal bandwidth of the amplifier used, but not too close to the frequency of the power-line interference.

The calculations in this section are made for a sinusoidal voltage V_{add} with an amplitude of 10 mV and a frequency of 1.2 kHz. In this example the frequency was chosen relatively high in order to be able to observe

3.3. Method

the variation of the electrode-skin impedance imbalance resulting from manually pressing electrodes to the skin. The imbalance variation resulting from manual movements have an upper frequency limit of about 30 Hz.

In order to quantify the electrode-skin impedance mismatch of a pair of electrodes we measure the amplitude and phase of the known common-mode signal after the first amplifier stage. This is done by synchronous demodulation, e.g., by multiplying the signal of interest with V_{add} followed by a lowpass filter (this is the same principle that lock-in amplifiers use).

The block diagram of the amplifier for bioelectric events is depicted in Fig. 3.1.

The above figure represents a generalized view of the schematic which can be adapted from one to 'n' channels. Each channel is preceded by a protection circuit which limits the maximal current through the body to 50 μA_{rms}.

For our test application we built the instrumentation amplifier (INA) in Fig. 3.1 using three low-power CMOS op-amps of the type LMC6462 (National Semiconductor) chosen for its high input resistance ($R_{in} > 10^{12} \Omega$, $C_{in} = 1.5$ pF) . The gain of the INA was set to 20 dB. The CMRR of the INA was maximized by manually trimming one resistor [Pall 91]. For the DRL circuit we used the TS914 (STMicroelectronics) chosen for its larger bandwidth.

The measurement circuit operates as follows: any voltage appearing at the positive input of the DRL op-amp OP_{DRL} forces the DRL loop to react in such a way that the differential input voltage of the DRL op-amp is again zero (principle of virtual ground). For a sinusoidal signal this is true as long as the frequency remains within the bandwidth of the DRL loop. As a result, the sinusoidal voltage V_{add} appears as a common-mode voltage to the input of the instrumentation amplifier (INA). This known common-mode voltage is rejected by the CMRR of the INA, except for the part which is converted to a differential signal by the electrode-skin impedance mismatch (this is known as the potential-divider effect, see section 2.2.4). The signal is then amplified by the differential-mode gain of each INA and appears superimposed on the corresponding output voltage $V_{out\ i}$. If we ignore all other sources, e.g., power-line interference and bioelectric signal, we can calculate the residue of the known common-mode signal $V_{add\ i}$ at the electrode 'i'.

This residual voltage is:

$$V_{add\ i} = \frac{1}{\text{CMRR}_{pot}} \cdot G_{DM} \cdot V_{add} \qquad (3.2)$$

3. Monitoring Electrode-Skin Impedance Mismatch

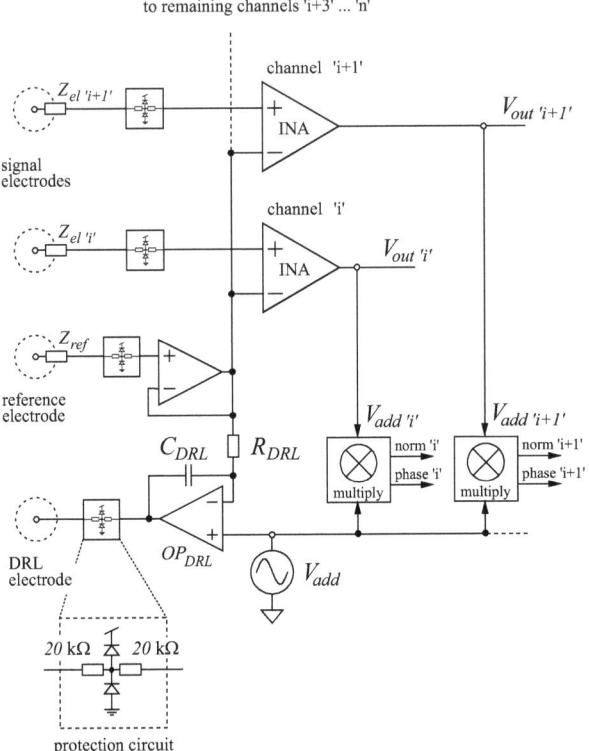

Fig. 3.1: Block diagram of the amplifier for bioelectric events. The amplifier has one reference electrode, one DRL (driven-right-leg) electrode and 'n' signal electrodes of which only two are shown. Each channel has one instrumentation amplifier (INA) with a gain of 20 dB. A known common-mode voltage V_{add} is applied to the positive input of the op-amp OP_{DRL} which is part of the DRL loop.

This equation is valid as long as the CMRR of the INA is much larger than $CMRR_{pot}$. $CMRR_{pot}$ stands for the common mode rejection ratio resulting from the potential-divider effect. The corresponding formula is derived in section 2.2.4):

$$\mathrm{CMRR}_{pot} = \frac{\overline{Z_{in}}}{|\Delta Z_{el}|} \qquad (3.3)$$

ΔZ_{el} stands for the difference of two electrode-skin impedances and $\overline{Z_{in}}$

3.3. Method

represents the mean value of the two common mode input impedances of the amplifier. G_{DM} represents the differential-mode gain (for the frequency of V_{add}).

Combining equation (3.3) and (3.2) yields:

$$V_{add\ i} \approx \frac{Z_{el\ i} - Z_{ref}}{\overline{Z_{in}}} G_{DM} \cdot V_{add} \qquad (3.4)$$

In this particular application and for the chosen frequency of the known common-mode signal V_{add} the input impedance of the INA is primarily capacitive. With this in mind, we can solve for the input electrode-skin impedance mismatch:

$$Z_{el\ i} - Z_{ref} \approx \frac{1}{j\omega \overline{C_{in}} G_{DM}} \frac{V_{add\ i}}{V_{add}} \qquad (3.5)$$

In the measurement circuit we quantify the residual voltage $V_{add\ i}$ by a multiplication with the original common-mode signal V_{add}. This is usually called synchronous demodulation and results in two signals 'norm i' and 'phase i' representing amplitude and phase of the residual signal. These two signals can then be used to calculate the electrode-skin impedance mismatch between the reference electrode and the signal electrode 'i' according to equation (3.5).

In a real measurement situation $V_{out\ i}$ is composed of the amplified bioelectric signal and the residue of the known common-mode voltage V_{add}. To improve rejection of the bioelectric signal either a bandpass filter (here at 1.2 kHz) can be used or a longer integration period of the two signals 'norm i' and 'phase i'.

To measure the electrode-skin impedance mismatch between two signal electrodes i and j, the corresponding residue $V_{out\ i}$ and $V_{out\ j}$ are measured separately and then subtracted mathematically. For an absolute (precise) measurement of the electrode-skin impedance mismatch the mean input capacitance $\overline{C_{in}}$ of the INAs must be known very precisely along with the common-mode gain G_{DM}. If these values are not known it is still possible to calibrate the circuit using a resistor with a known value.

3.3.1 Experimental Results

To validate the method we quantify the amplitude of the residual common-mode voltage $V_{add\ i}$ using known impedances and a sinusoidal voltage V_{add} with an amplitude of 10 mV and a frequency of 1.2 kHz.

3. Monitoring Electrode-Skin Impedance Mismatch

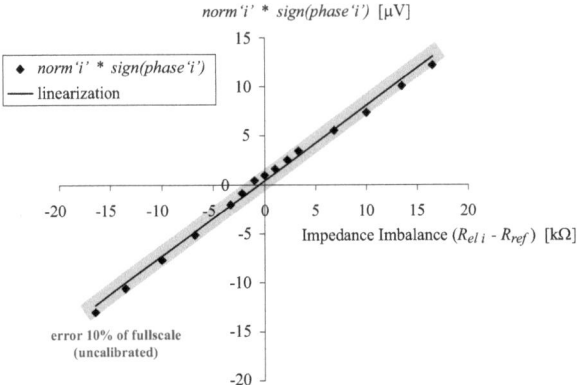

Fig. 3.2: The norm of the residue of the common-mode signal $V_{add\,i}$ multiplied by the corresponding phase as a function of the artificial impedance mismatch $R_{el\,i} - R_{ref}$ added to an ECG phantom.

Fig. 3.2 shows the measurements which have been carried out using an ECG phantom and a set of known resistors which were added in series to the input of the amplifier for bioelectric events. Looking closely at equation (3.5) and knowing that the difference between to resistors is a real number, we deduce that $V_{add\,i}/V_{add}$ results in an imaginary number. As a consequence the phase shift between $V_{add\,i}$ and V_{add} assumes mostly two distinct values, i.e., $\pm 90°$. The sign of $V_{add\,i}/V_{add}$ corresponds to the sign of $Z_{el\,i} - Z_{ref}$ and indicates which of the two involved resistances is larger.

The residual voltage is measured both in amplitude and phase. In Fig. 3.2 the amplitude (norm) is drawn while the phase is used to determine the sign. As predicted by equation (3.5) the phase shift was either $+90°$ (positive sign) or $-90°$ (negative sign).

Fig. 3.2 confirms the linear relation between the residual voltage expressed by $norm\,i \cdot sign(phase\,i)$ and the resistive mismatch $R_{el\,i} - R_{ref}$. From the figure the gradient of the linearization can be estimated to about $0.75\ \mu V/k\Omega$. The gradient can also be predicted from equation (3.5), using the values from our circuit ($V_{add} = 10$ mV, $C_{in} \approx 1.5$ pF, $f = 1.2$ kHz, $G_{DM} = 20$ dB) the resulting gradient would be $1.1\ \mu V/k\Omega$ which is close to the measured value. [1]

---

[1] The larger the CMRR the smaller the residual voltage. For the measurements described in the publication we added two switched-capacitor bandpass filters with a gain of 40 dB each to $V_{add\,i}$ and V_{add}. This was done prior to the synchronous demodulation and resulted in a stable measurement in spite of the small residual voltage.

3.3. Method

The absolute values of $R_{el\,i}$ and R_{ref} have no influence on the residual voltage.

Having gauged the measurement signal with known resistances we can now estimate the electrode-skin impedance mismatch when measuring a bioelectric signal. Without calibration the error is about 10 % of the shown scale and 1 % of the overall full-scale value (200 kΩ). The linearity can be improved by increasing the amplitude of the additional common-mode signal V_{add} while reducing both the CMR (common mode range) and the maximal gain of the amplifier.

Calibration is only required if the absolute value of the electrode-skin impedance mismatch is to be measured, because the impedance Z_{in} of the INA may vary from device to device. Calibration is not required when only the power-line vulnerability condition (i.e., an estimation of CMRR_{pot}) is to be determined according to equation (3.2).

Electrode-Skin Impedance Mismatch versus Applied Force

We now apply a sinusoidal force to the signal electrode 'i' and both the norm of the residual voltage $|V_{add\,i}|$ and the bioelectric recording are measured. For this measurement, a subject is lying on the ground with three dry electrodes (custom made Ag/AgCl electrode with a diameter of 12 mm attached to a dry gel pad with a diameter of 30 mm) pressed to the body by individual weights of 300 g. One electrode is then pulled away from the body with a sinusoidal force corresponding to a weight of 100 g and a frequency of 0.33 Hz. Fig. 3.3 shows the corresponding ECG measurement.

As a result of the sinusoidal force, the bioelectric signal displays a sinusoidal disturbance of the same frequency, i.e., 0.33 Hz. The norm of the residual voltage is digitally filtered by a lowpass filter with a corner frequency of 2.5 Hz and zero phase shift. The estimated electrode-skin impedance mismatch is also indicated with a second axis using the previously determined conversion factor 0.75 μV/kΩ. All signals are filtered by an additional digital 50-Hz notch filter for better visualization.

Fig. 3.3 demonstrates that there is a strong relationship between the measured electrode-skin impedance mismatch and the baseline variation. This is in accordance with previously published results [Hami 00], [Devl 84] showing a similar relation between impedance mismatch and baseline shift. This kind of artifact is called *skin-stretch artifact* or *skin-motion artifact* [Sear 00].

The figure in the publication showed the real measurement while the gradient was given based on the circuit described which had only one switched-capacitor bandpass filter and consequently did not match the gradient in the corresponding figure.

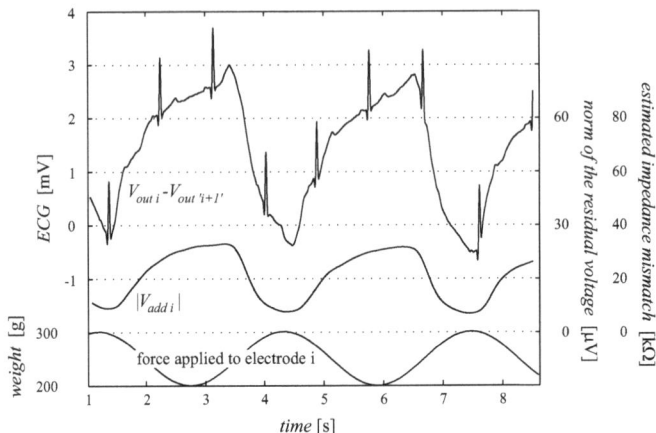

Fig. 3.3: A simultaneous measurement of an ECG and the norm of the residual voltage $|V_{add\,i}|$ resulting from a sinusoidal force applied to electrode 'i'. The measurement reveals a strong relation between the norm of the residual voltage and the baseline variation of the ECG. The residual voltage can be used to estimate the input electrode-skin impedance mismatch.

However, the amplifier in this work has a very high input resistance ($R_{in} > 10$ TΩ) when compared to previous amplifiers. As a result there are nearly no additional currents through the electrode-skin interface at the frequency of interest (0.33 Hz). This is important because the amplitude of the motion artifact voltage is proportional to the current through the electrodes [Zipp 79b].

For this particular measurement an impedance variation of about 20 kΩ results in a baseline shift of about 3 mV. The measured impedance, however, is not sinusoidal. We believe this is a result of the previously reported nonlinear relationship between stretching and relaxation of the skin [Oedm 81] which is probably due to the viscoelastic properties of the skin.

Electrode-Skin Impedance Mismatch versus Series Resistance

An additional measurement on a test subject has been performed in order to verify the influence of the impedance mismatch on the bioelectric recording. This time a 20 kΩ potentiometer is added in series to the input of the signal electrode 'i' and the resistance of the potentiometer is altered manually. The corresponding measurement is shown in Fig. 3.4.

3.3. Method

Again, a digital 50-Hz notch filter is used for better visualization.

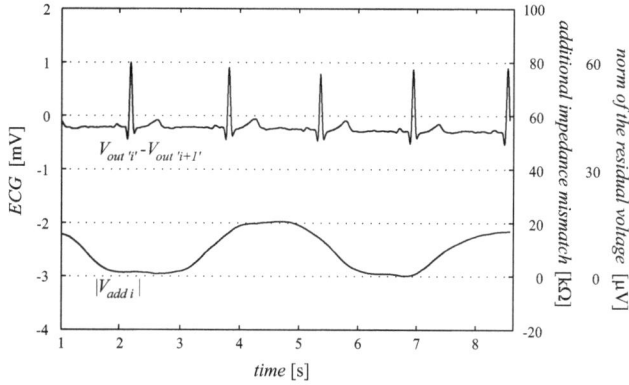

Fig. 3.4: Another simultaneous measurement of an ECG and the norm of the residual voltage $|V_{add\,i}|$ generated by manually varying a 20 kΩ potentiometer in series to the input of the electrode 'i'. Although the variation of the resistance is clearly visible, there is no baseline variation at all.

This measurement indicates that the prior observed baseline variation is not a result of the current flowing into the electrodes, i.e., the baseline variation is not a result of the potential-divider effect. The finding is not unexpected. In order to have an impedance variation of about 20 kΩ result in a voltage change of about 3 mV, a current of close to 150 nA must flow. The bias current of the INA is only in the order of some pA. Even when considering that the impedance variation is measured at 1.2 kHz and the baseline variation takes place at 0.33 Hz, the bias current of the INA can be neglected.

But then, how does the measured electrode-skin impedance mismatch relate to the observed baseline variation? We propose that there are two distinct effects: a change of the resistance and a change of the capacitance within the skin. By compressing the skin, the resistance of the skin varies. If we consider the resistivity of the skin ρ constant (as a first approximation) we can describe the resistance of the skin by:

$$R_{skin} = \rho \frac{d}{A} \qquad (3.6)$$

Where A and d denote surface and thickness of the skin region under investigation, e.g., the skin under the electrode.

Pressing an electrode to the skin will reduce the thickness of the skin and therefore result in a change of the electrode-skin impedance imbalance measured by our method.

The baseline variation is produced by the second effect, the variation of the capacitance built by inner and outer layer of the skin. Between these two layers there is a voltage difference of about 100 mV which is maintained by the body to form a protective barrier. This voltage is maintained by a ionic current, i.e., the body is actively transporting ions through the skin, e.g., via the sweat glands. In [Talh 96] this current is called 'injury current'. The 'injury current is balanced by the drift current i_d resulting from the electric field across the capacitor (i.e., the skin).

$$i_d = V_{skin} \frac{A}{\rho d} \qquad (3.7)$$

In the presence of an electrode this balance is disturbed by the bias current of the amplifier. The bias current i_b adds to the 'injury current'. The proposed model is depicted in Fig. 3.5:

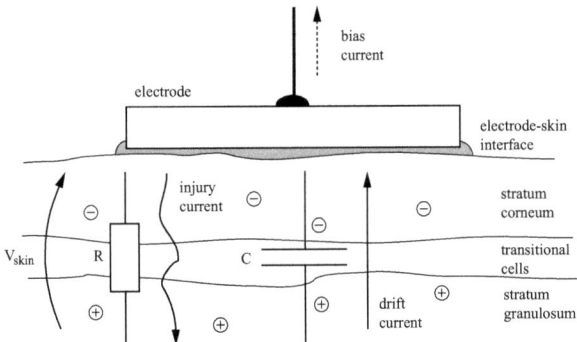

Fig. 3.5: The body generates a negative voltage between the inner and outer skin by an ionic current called by some authors 'injury current'. The current is mainly balanced by a drift current. The different skin-layers act like a capacitor.

The three currents are in equilibrium:

$$i_d = i_{injury} + i_{bias} \qquad (3.8)$$

Applying a force to the electrode will disturb the balance of the three currents. Compressing the skin immediately varies the capacitance of the capacitor built by the different skin-layers, resulting in a voltage change over the capacitor. Likewise, the inter-skin resistance of the skin decreases and thus the injury current generates a lower voltage drop. These two fast effects are followed by a slower process during which a new equilibrium is reached. The new equilibrium is defined by the new capacitance, the new inter-skin impedance and the three currents.

This model is in agreement with the characteristics of motion artifacts reported in the literature. The model also proposes a fast and slow effect as described in [Oedm 81]. The new equilibrium is characterized by a larger capacitance. Accordingly, starting from a compressed skin will result in a larger time constant (until a new equilibrium is reached) than starting from a non-compressed skin as it was described in [Talh 96]. Our model also anticipates that removing the upper part of the skin reduces motion artifacts [Burb 78] because part of the inter-skin capacitor is destroyed and the stored charge reduced.

3.4 Novelty

The method discussed above is the first method which allows to measure the electrode-skin imbalance concurrently with the bioelectric signal without altering the input impedance of the amplifier for bioelectric events.

This method can also be used to give real-time feedback to an operator on the quality of the electrode-skin imbalance. It is possible to detect the lift-off of one electrode. The method is also capable of generating an error signal for adaptive filtering of motion artifacts.

Based on the measurement signals, we proposed the first skin model for motion artifacts comprising a variable capacitance. The skin model has the potential to explain many of the reported characteristics of motion artifacts.

4. LOW-NOISE TWO-WIRED BUFFER ELECTRODES FOR BIOELECTRIC AMPLIFIERS

This chapter is a synopsis of the corresponding publication in the IEEE Transactions on Biomedical Engineering [Dege 07].

4.1 Problem Statement

Bioelectric recordings suffer from extensive power-line interference and other artifacts due to the high source impedance of the biological signal and the long wires, notably in the case of EEG measurements.

It has been proven many times that buffering the signal using active electrodes reduces the interferences. Despite the improvement of the signal quality, active electrodes are rarely used in practice. We attribute this to the noise added to the bioelectric signal and to increased complexity due to the higher number of connecting wires. The following work describes buffer electrodes with low noise and a low number of wires.

4.2 Prior Art

Active electrodes have been reported as early as 1968 when a single FET-transistor was used to buffer the signals from an ECG [Rich 68]. Several publications describe the design and use of buffer electrodes and their ability to reduce motion artifacts and render skin preparation unnecessary while improving the quality of the signal in general [Nish 92], [Ko 98]. Most buffer electrodes reported in the literature are built using operational amplifiers (op-amps) in a voltage-follower configuration.

There is only one publication known to us which describes a two-wired buffer electrode. The electrode employs a custom-made integrated circuit [Padm 90], which, according the the circuit diagram, is based on a (customized) op-amp. It should be noted that the reported two-wired electrode has one particularity: the bioelectric signal has to be the lowest voltage in the circuit because the substrate of the circuit is used as the input node. This limits the number of electrodes to one electrode per integrated circuit. A circuit capable of driving these electrodes in respect of their particular requirement is not described in the publication.

There is one company known to us which markets two-wired buffered electrodes [bios]. Unfortunately, the design of their electrodes "active two" is not disclosed to the public.

4.2.1 Pilot Study

To verify the claim that buffer electrodes reduce power-line interference [Fern 97], we built buffer electrodes using a CMOS amplifier in a standard buffer configuration employing three wires per electrode. The used op-amp was designed by us as part of an exercise to build integrated electrodes (not part of this thesis). To achieve minimal size, we bonded the op-amp directly onto a PCB (printed circuit board) with gold contacts (necessary for bonding). This results in the smallest-active-surface electrodes reported so far.

Fig. 4.1 details the different steps from fabrication to the application of the active buffer electrode. ① shows the naked PCB with the square pad, on which the chip (as a naked die) is glued. Also visible are some small squares beside the large square to which the connecting pads will be bonded. Once the chip is bonded to the PCB, the chip is covered by a glob top, and a single op-amp is cut off from the remaining PCB ②. On one side an Ag/AgCl cup electrode is soldered onto the PCB ③ and the overhanging corners are cut off. To the other end of the PCB a very thin and flexible three-lead wire ④ is soldered (no shielding is required). The whole PCB is drenched in epoxy and covered by a shrink tube ⑤ to seal off the circuit from the environment.

To be able to add the buffer electrodes to an existing amplifier, e.g., in a hospital, we fabricated a very simple connector box (depicted in Fig. 4.1) comprising a battery and a resistive divider ⑥. Fig. 4.2 illustrates the schematic of the box with one buffer electrode drawn.

Three active buffer electrodes ⑦ made with tin-cup electrodes ⑤ were tested at the USZ (university hospital in Zurich) by recording an EEG. The connector box allows connecting the buffered bioelectric signal to the standard measurement box by a touch-proof connector ⑧. Fig. 4.3 shows the FFT (fast Fourier transform) of two signals recorded simultaneously at two close sites. The FFT on the left belongs to a standard

4.2. Prior Art

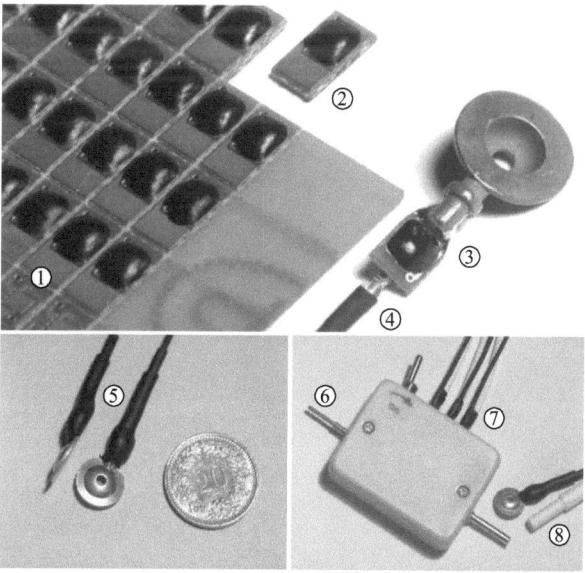

Fig. 4.1: Smallest active electrode to date made by directly gluing the operational amplifier to the board, bonding the contacts and protecting the circuit with a glob top. Numbers are explained in the text.

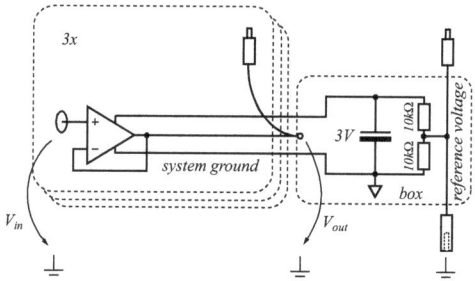

Fig. 4.2: The schematics of the buffer electrodes and the corresponding driver circuit. The latter consisting of nothing more than a battery and a resistive divider. The resistive divider is used to connect the external reference voltage (from the existing amplifier) to the mid-supply of the buffer electrodes.

passive Ag/AgCl electrode, the one on the right to our active buffer electrode.

64   4. Low-Noise Two-wired Buffer Electrodes

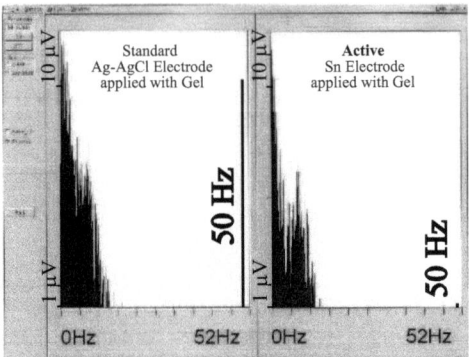

Fig. 4.3: The FFT of two EEG signals recorded with a passive and an active electrode at two close sites. The FFTs correspond to a signal stretch of about 2 s sampled with 416 Hz, only the part from 0 to 52 Hz is shown.

The two FFTs look similar except for the 50-Hz component, which is about 10 times larger for the passive electrode than for the active electrode. This proves that the use of active electrodes reduces the power-line interference. The improvement corresponds to about 18 dB (@ 50 Hz). Both electrodes were connected to the same amplifier by means of unshielded leads. The recordings were taken in a shielded (!) room at the USZ.

The pilot study proved to us the beneficial effect of buffer electrodes, but also showed us the disadvantage of using electrodes with three wires, i.e., the increased complexity when connecting to the circuit and the increased stiffness of the cable. The aim was to develop buffer electrodes with two wires and a low noise contribution.

4.3 Method

To reduce the number of wires, we designed active electrodes using a current supply and voltage output. To design the electrode with as low noise as possible, we reduced the number of active elements to the minimum: one transistor (MOS-FET) as input buffer and one transistor (J-FET) for the current-source. For comparison we also designed a two-wired buffer electrode using an off-the-shelf op-amp.

Amplifiers for bioelectric events require a high input impedance. For this reason we use a FET transistor for the single-transistor electrode and a CMOS op-amp for the single-op-amp electrode. The two buffer

4.3. Method

electrodes are drawn in Fig. 4.4:

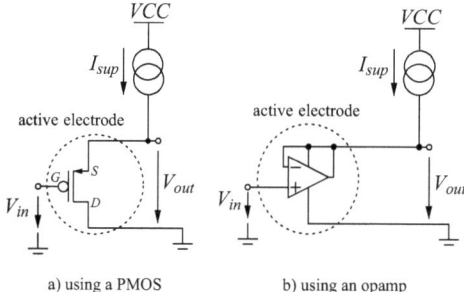

a) using a PMOS   b) using an opamp

Fig. 4.4: Principle of an active buffer electrode biased by a current-source with either a) an FET transistor or b) an CMOS op-amp employed as a voltage follower.

Both buffer electrodes are implemented as voltage followers. Fig. 4.4 a) shows a P-MOS-based level shifter using a single FET (NDS336 from Fairchild Semiconductor). This P-MOS transistor was chosen because of its low threshold voltage and a very small channel-length modulation parameter λ. Fig. 4.4 b) depicts an op-amp-based buffer electrode using the LMC7111 (National Semiconductor). This CMOS-input op-amp was chosen because of the low supply voltage of 2.7 V.

The two buffer electrodes were supplied by a current-source of the type J505 (Vishay) which consists of a single J-FET (Junction FET). The J505 was chosen for its simple topology. To source 1 mA the J505 requires a voltage drop of about 1.2 V. Fig. 4.5 shows an extract of the corresponding datasheet.

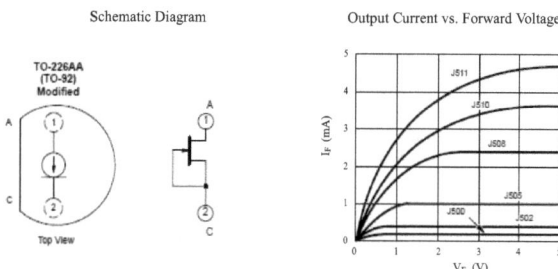

Fig. 4.5: An extract of the datasheet of the J505 current-source from Vishay showing the schematic of the current-source as well as the graph of the forward current I_F in function of the forward voltage V_f.

The output voltage V_{out} of both electrodes is identical to the supply voltage of the device and can be written as:

$$V_{out\ MOS} = -V_{DS} = V_{in} - V_{GS} \qquad (4.1)$$

$$V_{out\ opamp} = \frac{A_0}{1+A_0} V_{in} \qquad (4.2)$$

Where $V_{GS} < 0$ denotes the gate-source voltage of the P-MOS device and A_0 stands for the open-loop gain of the op-amp. It should be noted that in the case of the MOS-based electrode the gate-source voltage V_{GS} adds an offset to the output voltage $V_{out\ MOS}$. The value of V_{GS} depends on the technology used but also on the supply current I_{sup} and is normally in the order of several hundred millivolts:

$$V_{GS} = \sqrt{\frac{2I_{DS}}{\beta(1+\lambda V_{DS})}} + V_{th} \qquad (4.3)$$

Equation (4.3) is valid for a MOS transistor in saturation. $I_{sup} = -I_{DS}$ stands for the supply current of the buffer electrodes, β for the gain factor (technology dependent) and λ denotes the channel-length modulation parameter. $V_{th} < 0$ is the threshold voltage of the chosen P-MOS transistor and can vary between identical devices by ± 100 mV. For bioelectric signals this offset will be added to the offset from the electrode-skin interface. Because amplifiers for bioelectric events need to be able to handle offset values as large as ± 300 mV, it is reasonable to assume that this additional offset can be handled too and may only require a little adjustment in the design of the amplifier.

Note, the op-amp of the buffer electrode in Fig. 4.4 b) is not able to actively drive the output voltage $V_{out\ opamp}$, the correct output voltage can only be reached because of the current-source. If, for example, the input voltage V_{in} raises above the output voltage $V_{out\ opamp}$ (which corresponds to the supply voltage of the op-amp) the output stage of the op-amp reacts by reducing the impedance between the positive supply and the output. This has no effect as the two terminals are already short-circuited. However, the resistance between the output and the negative supply of the op-amp increases also and the constant current from the source will therefore cause the output voltage to rise until the output voltage matches the input voltage again. The current-source (J-FET) can be considered to be the upper part of the output stage of the buffer electrode.

The requirements for the op-amp are that one supply is included in the input range and that the maximal current the output stage of the op-amp can handle being connected to this supply exceeds the value of the supply current I_{sup}. The active electrode with an op-amp has already been described, although using an custom-made integrated circuit [Padm 90].

4.3. Method

The gain of the two active electrodes can be shown to be close to unity:

$$\text{Gain}_{MOS} = \left(\frac{\partial V_{in}}{\partial V_{out}}\right)^{-1} = \left(1 - \frac{\partial V_{GS}}{\partial V_{DS}}\right)^{-1} = \left(1 - \frac{\partial V_{GS}}{\partial I_{DS}} \cdot \frac{\partial I_{DS}}{\partial V_{DS}}\right)^{-1}$$
$$\approx 1 - \frac{1}{g_m r_0} \tag{4.4}$$

with

$$g_m = \frac{2I_{DS}}{V_{GS} - V_{th}} \quad \text{and} \quad r_0 = \frac{1 + \lambda V_{DS}}{\lambda I_{DS}} \tag{4.5}$$

$$\text{Gain}_{opamp} \approx 1 - \frac{1}{A_0} \tag{4.6}$$

Where g_m denotes the forward transconductance and r_0 the output resistance of the MOS device. Equation (4.4) is valid for a MOS transistor in saturation.

The next important characteristic of a buffer electrode is the output resistance. The output resistance should be low in order to reduce the interference when long wires are used, e.g., for EEG applications. The output resistance of the op-amp-based electrode is very low because of the negative feedback and is below some Ω. The output resistance of the P-MOS-based electrode is given by:

$$R_{out\,MOS} = \frac{\partial V_{out}}{\partial I_{sup}} = \left(\frac{\partial I_{DS}}{\partial V_{DS}}\right)^{-1} \quad \text{with} \quad V_{GS} = V_{DS} + V_{in} \tag{4.7}$$

$$= \left(\beta(V_{GS} - V_{th})(1 - \lambda V_{DS}) + \frac{\beta}{2}(V_{GS} - V_{th})^2 \lambda\right)^{-1} \tag{4.8}$$

$$= \left(g_m + \frac{1}{r_0}\right)^{-1} = \frac{1}{g_m + \frac{1}{r_0}} = \frac{1}{\frac{2I_{DS}}{V_{GS}-V_{th}} + \frac{\lambda I_{DS}}{1+\lambda V_{DS}}} \tag{4.9}$$

for $\lambda \to 0$

$$R_{out\,MOS} \approx \sqrt{\frac{1}{2\beta I_{sup}}} \tag{4.10}$$

Again we see that the output resistance depends on the supply current I_{sup}. The higher the supply current the lower the output resistance $R_{out\,MOS}$ of the MOS-based electrode.

The most important characteristic for a buffer electrode is the spectral voltage-noise density of the electrode, because the noise of the electrode is added one-to-one to the bioelectric signal. The spectral voltage-noise density of the op-amp can be found in the datasheet. The spectral voltage-noise density of the transistor is a function of the supply current. For example, the square value of the spectral voltage-noise density for frequencies above the flicker noise corner frequency can be expressed by:

$$\overline{v}_{n\ MOS}^2 \approx \frac{4k_BT\alpha}{g_m} = \frac{4k_BT\alpha}{\sqrt{2\beta I_{sup}(1+\lambda V_{DS})}} \qquad (4.11)$$

Where k_B stands for the Boltzmann constant, T for the temperature in Kelvin and α for the transistor noise coefficient (a typical value for α is 2/3).

We can now estimate the required supply current I_{sup} for the two different active buffer electrodes. For the op-amp-based electrode we recommend to use at least twice the supply current of the op-amp as a supply for the electrode to provide for any tolerances. For the MOS-based electrode we have a lower limit given either by the desired noise level, which then has to be matched to equation (4.11), or by the maximal desired output resistance, which is given by equation (4.10). A low supply current will also reduce the bandwidth of the P-MOS-based electrode, which could eventually lead to a reduction of the CMRR due to the frequency mismatch between different electrodes, similar to the effect discussed in section 2.2.4.

The active buffer electrodes described above can be added to an existing amplifier for bioelectric events as long as an additional current-source is provided to supply the buffer electrodes. The electrodes themselves have a limited input CMR (common mode range) due to the voltage drop over the current-source and, for the op-amp-based electrode, the requirement for the minimal voltage supply of the op-amp.

For the MOS-based electrode (in saturation) the minimal output voltage $V_{out\ MOS}$ corresponds to $V_{th} - V_{GS}$ (this is the inverse of the overdrive voltage which depends on I_{sup} and can be several hundred millivolts), the minimal input voltage equals V_{th} which is about -780mV for the NDS336. The op-amp-based electrode requires a minimal output voltage $V_{out\ opamp}$ of 2.7 V (for the LMC 7111) which coincides with the minimal input voltage.

We therefore strongly recommend the use of a DRL circuit to drive the input common-mode voltage of the amplifier to the midrange voltage of the input CMR of the employed active buffer electrodes (see also section 2.3.3). The use of a DRL circuit to drive the DC value of an active electrode is demonstrated for the amplifier with two-wired amplifying electrodes in chapter 6.

We will resume the results for the chosen reference currents later in table 4.1.

4.3. Method

4.3.1 Results

We fabricated several prototypes of two-wired active buffer electrodes using buttons for trousers on which we soldered a small PCB. This design allowed us to attach different disposable electrodes. Fig.4.6 shows an Ag/AgCl brush-type EEG electrode clipped to one of the active electrodes.

Fig. 4.6: A shielded two-wire active electrode mounted on a button in order to attach different types of disposable electrodes.

The prototype electrodes are left unprotected to allow the soldered components to be easily changed.

We first measured the input-referred spectral voltage-noise density of the two different electrodes for a supply current I_{sup} of 1 mA. In Fig. 4.7 the spectral voltage-noise density of both two-wired active buffer electrodes is presented together with the spectral voltage-noise density of an Ag/AgCl electrode[1] for comparison.

The noise of the P-MOS is very similar to the noise of the Ag/AgCl electrode except for the low-frequency part. At low frequencies the noise of the P-MOS is higher, the corner frequency from which the spectral voltage-noise density shows a frequency dependency is higher for the P-MOS than for an Ag/AgCl electrode. We can estimate the total noise by approximating the area under the curves.

We then estimate the total voltage noise from 0.1 Hz to 200 Hz (ECG) to 2.05 μV_{rms} for the MOS-based electrode with a supply current of 1 mA. This is higher than the total voltage noise of an Ag/AgCL electrode that amounts to 1.4 μV_{rms}. The op-amp-based electrode does visibly have more noise at lower frequencies and does not present itself as a good solution for ECG.

---

[1] The measurements of the electrode noise were carried out by biosemi [bios], explanations at "http://www.biosemi.com/faq/without_paste.html"

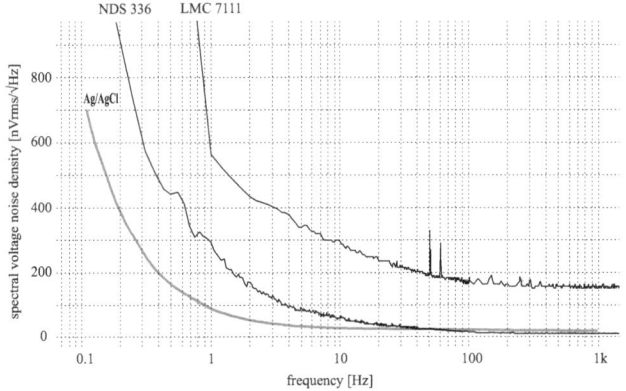

Fig. 4.7: The spectral voltage-noise density of the two different active buffer electrodes measured with a supply current of 1 mA using a vector signal analyzer (Stanford SR785). Different sampling rates were used above and below 100 Hz.

If we estimate the total voltage noise from 10 Hz to 1 kHz (EEG) we get 11.1 μV_{rms} for the op-amp-based electrode, which is nearly ten times higher than the total voltage noise of an Ag/AgCL electrode in this frequency band (1.5 μV_{rms}). A much better choice for EEG is the MOS-based electrode, which has a total voltage noise of less than 1 μV_{rms} for the same frequency band. This is even lower than the inherent noise of the Ag/AgCl electrode. We do not recommend reducing the supply current of the MOS-based electrode because this would result in more low-frequency noise for which the MOS-based electrode is worse than the Ag/AgCl electrode. With a supply current of 1 mA the spectral voltage-noise density of the electrode is higher than that of the MOS-based electrode for frequencies above about 100 Hz. The spectral voltage-noise density of the MOS-based electrode at 1 kHz is around 10 nV/$\sqrt{Hz}$ (this is the limit of the Stanford SR785).

As a first conclusion we can state that the total voltage noise of the op-amp-based electrode is more than ten times higher than the total voltage noise of the MOS-based electrode. As long as the additional DC offset of ±100 mV added by the MOS based active electrodes does not saturate the subsequent amplifier, the MOS-based electrodes will allow to measure signals which are ten times smaller. If the buffer electrode is to be added to an existing amplifier which cannot be adapted to adjust for the additional DC offset, the op-amp-based electrode may be used in spite of the larger noise contribution.

Table 4.1 shows the most important characteristics measured for the

4.3. Method

two buffer electrodes . The MOS-based electrode from Fig.4.4 a) was measured with two different reference currents, i.e., 1 mA and 10 μA for comparison.

Tab. 4.1: The measured characteristics for both active electrodes.

Components		P-MOS NDS336		op-amp LMC7111
I_{sup}	[mA]	1	0.01	1
Offset	[mV]	840 ± 25	620 ± 17	3 ± 1
VCC	[V]	$\geq$1.84 V	$\geq$1.62 V	$\geq$3.7 V
Noise[a]	[μV_{rms}]	0.96	1.95	11.1
Gain		0.99986	0.9999993	0.99986
R_{in}[b]	[$G\Omega$]	3000	3000	5000
R_{out}	[Ω]	50	4.1k	0.1
I_{bias}	[pA]	-0.060	-0.060	-31

[a] total noise from 10 Hz to 1 kHz
[b] Measured with a parameter analyzer HP4155B/4156B

To test the electrodes we recorded the ECG of the author with all three active buffer electrodes from Tab. 4.1. The amplifier used was an experimental set-up with a gain of 60 dB based on the INA122 (orig. Burr-Brown, now TI) without additional filtering. We measured an ECG because we estimate that the quality of an amplifier can be best evaluated with the distinct features of an ECG recording, e.g., the absence of noise and the flatness of the baseline.

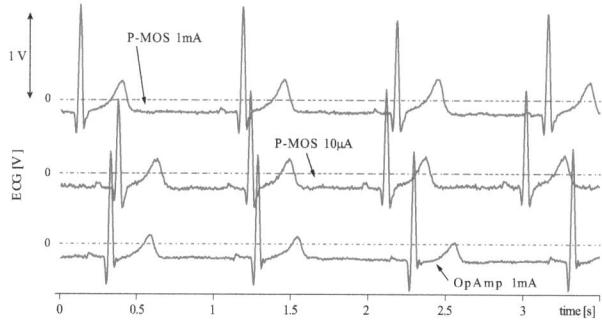

Fig. 4.8: Three ECG curves from the author recorded at different times. The type of the electrode as well as the supply current are included as part of the curve's name.

The recordings were taken on different days using disposable dry pre-gelled adhesive electrodes. No prior cleaning of the skin was necessary. Two recordings were taken with a P-MOS based two-wired active electrode but with different supply currents. One recording was taken with an op-amp-based electrode.

It should be noted that the trace with the lower reference current (P-MOS 10uA) appears to display slightly more noise.

4.4 Novelty

We have successfully demonstrated the use of a single transistor for an ECG recording using two-wires per electrode. Although the use of a single transistor was not new, it was somehow no longer considered because of the large prevalence of op-amps. We have shown that the use of a single transistor adds ten times less voltage noise to the bioelectric signal than the use of an op-amp (for the same supply current). In addition, the input bias current is minimal when using a single transistor which is important to reduce motion artifacts [Zipp 79b].

Furthermore, using a low-noise single-transistor-based active electrode ensures the maximal input impedance which is important for bioelectric recordings. Using a buffer electrode with a high input impedance allows us to use any type of amplifier for the subsequent stage, e.g., low-noise amplifiers with a bipolar input stage. This results in a much larger choice for the subsequent amplifier stage.

5. ENHANCING INTERFERENCE REJECTION OF AMPLIFYING ELECTRODES BY AUTOMATED GAIN ADAPTION

This chapter is a synopsis of the corresponding publication in the IEEE Transactions on Biomedical Engineering [Dege 04].

5.1 Problem Statement

In order to minimize the input-referred noise of an amplifier for bioelectric events, it is desirable to amplify the signal close to the source, i.e., already on the electrode. Yet, without any control of the gain, any gain mismatch between different electrodes will compromise the overall CMRR of the amplifier. The following work demonstrates an autonomous method to compensate for a gain mismatch of individual electrodes by adapting the gain of the subsequent differential amplifier.

5.2 Prior Art

A mismatch of the gain between different electrodes is known to be critical for achieving a good CMRR [Pall 91] (see also section 2.2.4). For example, when building a multi-electrode system with two-stage instrumentation amplifiers it is important that the amplifiers share a common reference node [Mett 94] ,[Mett 91].

This is because when using a common reference node the CMRR of the second stage is improved by the gain of the first stage as demonstrated in [Pall 91b].

Accordingly, amplifying electrodes have been demonstrated where the first stage of a two-stage amplifier with common reference node is placed on the electrodes. Yet, because the electrodes need a common reference voltage, the electrodes require up to five connecting wires [Valc 04].

In [Pall 91] it is also shown that for a single differential amplifier built using one op-amp and four resistors it is possible to improve the CMRR of the differential amplifier by manually trimming one of the four resistors.

In [Ober 82] a method was presented to estimate the electrode-skin impedance at the input of an amplifier for bioelectric events by forcing a known common-mode voltage via the DRL circuit.

5.3 Method

We combined the afore mentioned results and methods to compensate for the gain mismatch of amplifying electrodes by adjusting the gain of the following differential amplifier. The adjustment was done using a voltage-controlled resistor combined with synchronous demodulation all together being part of a PI-control loop.

The method was demonstrated with a known common-mode signal superimposed to the input via the DRL circuit as well as with the power-line interference at 50 Hz. Both times the common-mode signal holds the information on how to adjust the gain of the differential amplifier in order to improve the overall CMRR of the amplifier.

Using a known common-mode Signal

The first method we propose consists of applying a known common-mode signal V_{add} to the non-inverting input of a DRL circuit. We measure the amplification and phase shift of this known common-mode signal after the differential amplifier. These measurements will yield the control signals needed to implement a digital control loop which regulates the gain of the differential amplifier by means of a variable resistor.

By varying the voltage-controlled resistor we adapt the gain of the differential amplifier until the known common-mode signal V_{add} is minimized. This simultaneously maximizes the overall CMRR of the amplifier and minimizes the interference of any other common-mode signal, including the 50-Hz power-line interference. Fig. 5.1 shows the proposed circuit.

The DRL circuit was already introduced in section 2.3.3. As a short recapitulation we look at the known common-mode signal V_{add}. This signal will be amplified by the DRL circuit, applied to the body and then sensed by the two amplifying electrodes. The output voltages V_A and V_B of the two electrodes are averaged by a resistive divider built using two identical resistors R_0. The amplified common-mode voltage

5.3. Method

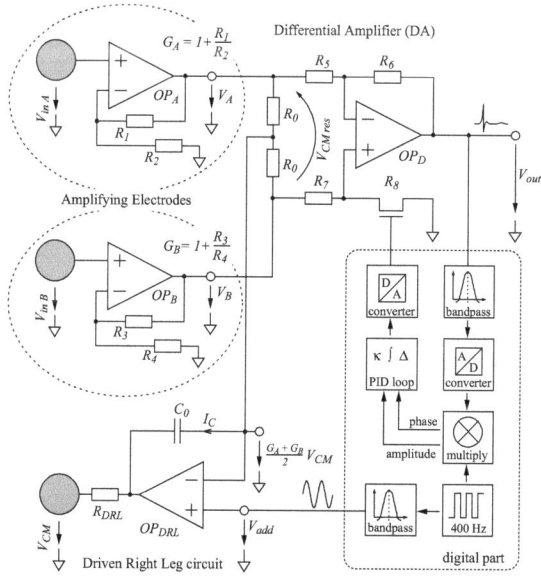

Fig. 5.1: A two-lead ECG amplifier composed of amplifying electrodes and a differential amplifier (DA) with adjustable gain and a digital control circuit to adapt the gain of the DA

will be equal to the known common-mode signal V_{add} as long as the DRL op-amp does not saturate (principle of virtual ground).

If there is a gain mismatch between the two amplifying electrodes, part of the common-mode signal will be converted into a differential-mode signal $V_{cm\ res}$ and amplified by the differential amplifier. The result of the gain mismatch is a reduced CMRR (see also section 2.2.4) leading to unwanted interferences. The residue of the common-mode signal $V_{CM\ res}$ can be expressed by:

$$V_{CM\ res} = (G_B - G_A)V_{CM} \qquad (5.1)$$

Where G_A and G_B stand for the individual gains of the two amplifying electrodes. The phase shift between the common-mode signal V_{CM} and the amplified residue $G_{DM} \cdot V_{CM\ res}$ at the output of the differential amplifier is then either $0°$ or $180°$. The aim of our method is to degenerate the CMRR of the differential amplifier in a controlled way such that the common mode after the differential amplifier and the amplified residue cancel each other out:

$$G_{DM} \cdot V_{CM\,res} = -G_{CM} \frac{G_A + G_B}{2} V_{CM} \qquad (5.2)$$

Where G_{DM} is the differential-mode gain and G_{CM} is the common-mode gain of the differential amplifier. Combining equation (5.1) and equation (5.2) yields the desired common-mode gain of the differential amplifier:

$$G_{CM} = -\frac{2(G_B - G_A)}{G_A + G_B} G_{DM} \qquad (5.3)$$

The controlled degeneration of the CMRR is achieved by varying the voltage-controlled resistor R_8 of the differential amplifier. We subsequently increase R_8 for a phase shift $>$ 90° and decrease R_8 for a phase shift $<$ 90° until we obtain a phase shift of exactly 90°.

It has to be noted that there might be an additional phase shift between the known common-mode voltage V_{add} and the common-mode voltage V_{CM} which is introduced by the DRL circuit (depending on the corner frequency $f_0 = 1/\pi C_0 R_0$ of the DRL circuit). When measuring the phase shift between the known common-mode voltage V_{add} and its residue this additional phase shift has to be taken into account (see section 5.3.1).

Using the 50-Hz Interference as common-mode Signal

Instead of adding a known common-mode signal, it would be sufficient to analyze any existing common-mode signal such as the 50-Hz power-line interference. The advantage of this method is obvious: no additional common-mode signal has to be added. The disadvantage is the small and arbitrary signal amplitude. The chosen common-mode signal needs to be amplified before processing.

Unlike the first method, no prior knowledge of the phase of the 50-Hz interference is available. Therefore, we measure the phase of the 50-Hz interference in reference to an internally generated reference signal (also 50 Hz) at *two* different locations in the circuit and determine the resulting phase shift by a simple subtraction of the two individual phase shifts. Fig. 5.2 depicts the circuit we designed:

The two locations used to measure the 50-Hz interference are the output voltage of the DRL circuit and the output voltage V_{out} of the differential amplifier.

Because the DRL circuit is an integrator, there is an additional phase shift of 90° between the common-mode signal at the output of the driven right leg and the amplified common-mode signal. Consequently, when using the power-line interference as common-mode signal, the desired

5.3. Method

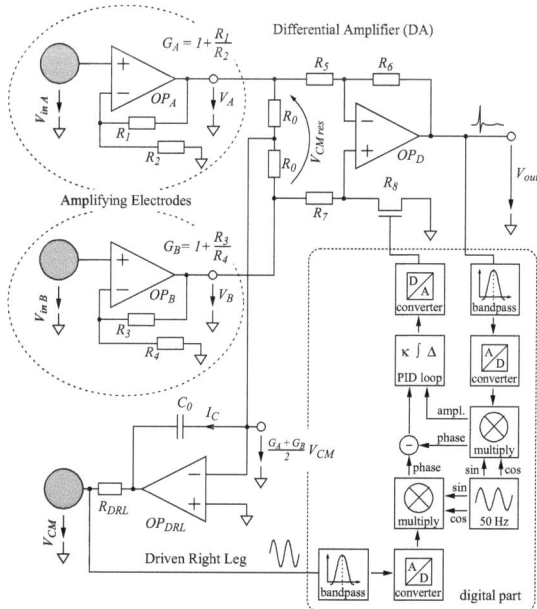

Fig. 5.2: A two-lead ECG amplifier where the gain of the differential amplifier (DA) is adapted until the phase shift of the 50-Hz power-line interference measured at different locations matches the desired value

phase shift is $0°$. We then increase R_8 for a phase shift $> 0°$ and decrease R_8 for a phase shift $< 0°$ until we obtain a phase shift of exactly $0°$. Again, depending on the corner frequency $f_0 = 1/\pi C_0 R_0$ of the DRL loop there may be an additional phase shift to be considered.

Both methods work well for signals which are below about 1 kHz. For higher frequencies the mismatch in the open-loop gain of the op-amps will determine the total CMRR and this cannot be compensated for by varying the resistor of the differential amplifier. For a very thorough analysis of the CMRR we refer to [Pall 91]. It is also important to note that the known common-mode signal V_{add} has to be amplified with the same gain (and phase) as the common-mode signal which should be minimized (e.g., the power-line interference) which normally means that the known common-mode signal V_{add} has to lie within the bandwidth of the amplifier.

5.3.1 Implementation

To vary the CMRR of the differential amplifier we built a voltage-controlled resistor using two FETs (field effect transistors) as shown in Fig. 5.3

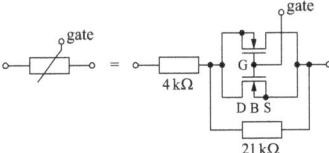

Fig. 5.3: The variable resistor as it was implemented, using two FETs in antiparallel configuration and two resistors for defining the minimal and maximal value of the resistor.

The resistance of this device is a function of the voltage over the gate. The minimal and maximal resistance is defined by two conventional resistors. For the implementation in Fig. 5.3 the two resistors are 4 kΩ and 21 kΩ, resulting in a minimal resistance of 4 kΩ (FETs fully on) and a maximal resistance of 25 kΩ (FETs fully off). The two FETs were used in an anti-parallel configuration to achieve symmetry. The remaining resistors of the DA have a value of 10 kΩ each. The resistance of the voltage-controlled resistor is not a linear function of the gate voltage.

It is important to ensure that the resistance of the voltage-controlled resistor R_8 does not change when the voltage drop over the resistor changes (e.g., in the presence of a ECG spike). Therefore, it is necessary that the two FETs are operated in the ohmic region, which is defined by the following basic inequality:

$$V_{DS} < V_{GS} - V_{th} \qquad (5.4)$$

V_{th} is the threshold voltage of the employed FETs and can be found in the datasheets. V_{GS} has best to be measured for the targeted resistance of the FET which in our case amounts to $V_{GS} = 1.4$ V for a target resistance of 16.8 kΩ. (The target resistance was chosen in a way to have a total resistance of 10 kΩ.)

In our circuit the maximum possible $V_{DS} = 6/40 \cdot$ VDD is achieved when the intermediate node V_B reaches VDD or VSS (assuming that the signal ground is at VDD/2). Assuming a threshold voltage V_{th} of 0.6 V the maximal supply voltage for this implementation is then 5 V. A higher supply voltage would eventually lead to the FETs not being operated in the ohmic region, as a result a change in V_{DS} would give rise to an undesired change in the resistance of the FETs.

5.3. Method

The control loop which is driving the gate voltage was implemented using a data-acquisition card (DAQ-1200 from National Instruments) in combination with a PC (running LabView). The AD converter on the DAQ card is an eight-channel 12-bit converter with 100kS/s. The DA converter is a two-channel 12-bit converter with 20kS/s. Bandpass filtering and anti-aliasing was added using switched-capacitor filters together with active analog filters. We implemented both methods separately.

Using a known common-mode Signal

We chose the known common-mode signal V_{add} to be a sinusoidal signal with a frequency of 400 Hz and an amplitude of 20 mV peak-to-peak. In this first example, we set the optimal phase shift to about $-45°$ by choosing an appropriate C_0. It is always possible to choose an appropriate C_0 for a common-mode signal of 400 Hz without violating the stability conditions for a DRL circuit [Wint 83]. It does, however, reduce the bandwidth of the DRL circuit. The reason we chose this particular value for C_0 was to simplify data processing. In section 5.3.1 we will present a method which will work with all values of C_0 (for the values used see table 5.3).

The voltage after the DA was first filtered by a switched-capacitor bandpass with a center frequency of 400 Hz and then sampled with 1.6 kHz. The center frequency of the bandpass is set by the same frequency which generates the 400-Hz common-mode signal. With this technique, a good agreement between the signal frequency and the bandpass is achieved.

To measure the phase shift, we use a very simple vector multiplication, which allows us to calculate the amplitude and the sign of the phase shift of the common-mode signal. I.e., through multiplication of the sampled values by the vector $\vec{v} = [1, -1, -1, 1]$, the resulting scalar, which we call error signal e, represents the component of the common-mode voltage having a phase shift of $+45°$ or $-135°$. This scalar e is used directly as an input to the controller. Fig. 5.4 clarifies the algorithm for three exemplary waveforms. Other waveforms may result in other values but over time the mean value will always be zero except for the desired waveform with the right frequency (principle of synchronous demodulation). Note, this particular algorithm is only valid for an optimal phase shift of $-45°$.

Again, the algorithm is based on the principle of synchronous demodulation. Even though other frequencies also contribute to the value of the multiplication, their contribution is canceled out by averaging. We take an average over 20 periods which we then feed to the digitally

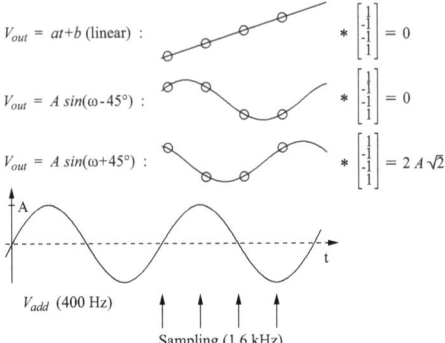

Fig. 5.4: V_{out} is sampled with 1.6 kHz and then multiplied with the vector $\vec{v} = [1, -1, -1, 1]$. Only components of the same frequency with a phase shift of either $45°$ or $-135°$ result in a constant scalar $\neq 0$.

implemented PID-controller, with 'P' standing for proportion, 'I' for integration, and 'D' for differentiation.

The controller is implemented using a recursive algorithm of the general form:

$$u(k) - u(k-1) = q_0 e(k) + q_1 e(k-1) + q_2 e(k-2) \tag{5.5}$$

$$q_0 = K\left(1 + \frac{T_D}{T_s}\right) \tag{5.6}$$

$$q_1 = -K\left(1 + 2\frac{T_D}{T_s} - \frac{T_s}{T_I}\right) \tag{5.7}$$

$$q_2 = K\frac{T_D}{T_s} \tag{5.8}$$

$u(k)$ is the kth value of the output of the PID-controller, which will be translated into a voltage by a DA converter and applied to the gates of the FETs emulating R_8. e is the error signal which, in this case, is the averaged value of the vector multiplication representing the amplitude of the known common-mode signal V_{add} after the DA. K stands for the gain, T_I for the time constant of the integration, and T_D for the time constant of the differentiation. T_s is the time between samples. Table 5.1 resumes the values used.

The values in table 5.1 represent a more specific case of the controller used. Because $T_D = 0$ the algorithm acts as a PI-controller.

5.3. Method

Tab. 5.1: Values used for the digital controller

K	=	0.0005
T_D	=	$0s$
T_I	=	$0.05s$
T_s	=	$0.0025s$

Using the 50-Hz Interference as common-mode Signal

In order to use the power-line interference to control the voltage-controlled resistor R_8 we need to sample two different signals, V_{out} and V_{CM}. Both signals are first conditioned by two identical switched-capacitor bandpass filters with a center frequency $f_0 = 50$ Hz and an anti-aliasing lowpass filter with a corner frequency $f_c = 2$ kHz. Together, the signal-conditioning circuits have an overall gain of 32 dB at 50 Hz. Both signals are sampled with 8 kHz and digitally multiplied with a sine and a cosine of 50 Hz (also sampled at 8 kHz). The scalar result of the two multiplications (sine and cosine) is converted into a complex number. The complex modulus represents the amplitude; the complex argument the phase shift of the 50-Hz component of the sampled signals with reference to the internally generated sine of 50 Hz. The values are averaged over 20 periods. The phase shift of the 50-Hz common-mode signal between V_{CM} and V_{out} is then simply the difference of the two complex arguments.

The phase shift is entered into the digital control circuit in order to control the voltage-controlled resistor R_8. Table 5.2 resumes the values used.

Tab. 5.2: Values used for the digital controller

K	=	0.0005
T_D	=	$0s$
T_I	=	$1s$
T_s	=	$0.02s$

Again, the results presented later in this work were all obtained using the above values resulting in a PI controller. This leads to a slower but more stable settling of the controller.

Calibration Procedure

The electronic circuitry of the signal processing may cause additional phase shift (or the DRL circuit may have a different phase shift than intended). This additional phase shift must be subtracted from the result of the vector multiplication.

The automatic calibration procedure consists of averaging the phase shift for the maximal value of R_8 and the minimal value of R_8, in our case for $R_8 = 4\ k\Omega$ and $R_8 = 25\ k\Omega$. We chose a minimum and maximum value of R_8 which is 2.5 times smaller or larger, respectively, than the target value of 10 kΩ. Using a wider range for R_8 may interfere with the requirement of the FETs being in the ohmic region. Using a narrower range may have an adverse effect on the calibration procedure. The averaged phase shift is then used as the optimal phase shift which has to be reached.

The calibration procedure is applied prior to the measurements and can be reapplied when necessary, e.g., after a displacement of the electrodes or in regular intervals. During the calibration procedure no measurement is possible.

5.3.2 Results

Amplifying Electrodes

In order to evaluate the two proposed methods we designed amplifying electrodes using surface mounted devices (SMD). The op-amp used was a TS931 (SGS-Thomson). This is a standard low-power rail-to-rail CMOS op-amp in a tiny SOT23-5 package. The op-amp was chosen for its small form factor and its high input resistance. The intended gain was 29.8 dB. The PCB with the op-amp and two gain-setting resistors was either soldered to an Ag/AgCl cup electrode or a gold clip to allow disposable electrodes to be used. Fig. 5.5 depicts two different amplifying electrodes and the active circuit on PCB employed for both electrodes.

Using an Additional common-mode Signal

We tested the method by building the proposed circuit of Fig. 5.1 on a test board using standard components. The ECG signals were measured by a set of amplifying electrodes as depicted in Fig. 5.5. Dry-gel electrodes were attached to the gold-plated buttons and applied manually to the body. The third electrode was located on the left leg. All measurement equipment was powered by batteries. Table 5.3 lists the resistor values used:

The ECG was measured with two active electrodes having different gains of 30.9 dB (35.08) and 28 dB (25.12) respectively. This reduces the CMRR of the first stage to 9.6 dB as calculated in equation (2.23).

5.3. Method

Fig. 5.5: A pair of active electrodes and the designed PCB with SMD parts. The PCB is either soldered to an Ag/AgCl cup electrode or a gold clip used for disposable electrodes.

Tab. 5.3: Values used for our circuit

R_1, R_3	=	$680k\Omega$	R_2, R_4	=	$22k\Omega$ $(27k\Omega)$
R_5, R_7	=	$10k\Omega$	R_6, R_8	=	$10k\Omega$
R_0	=	$100k\Omega$	C_0	=	$10nF$

As a result of this very low CMRR the recording will be affected by large power-line interferences.

Fig. 5.6 shows an ECG recorded simultaneously with two different differential amplifiers (DA), each one followed by a lowpass filter with a gain of 29.5 dB. Both DA use the same amplifying electrodes. The standard DA is built with four resistors of 10 kΩ with a tolerance of 1 %. The second DA is built with the same set of resistors, but with the resistor R_8 being emulated by two FETs and two resistors as depicted in Fig. 5.3. The measurement is made using the first method based on a known common-mode signal V_{add} with an amplitude of 20 mV peak-to-peak and a frequency of 400 Hz. The waveforms in Fig. 5.6 were recorded while the FET transistors were regulated by the PI control loop. The initial settling time was several 100 ms.

An offset of +0.2 or −0.2 mV was added to the signals for better visualization. As expected, the ECG signal using a standard DA shows a large amount of 50-Hz interference. This is due to the low CMRR_{el}, which is a result of the difference in gain of the two amplifying electrodes. At the same time, the regulated DA is able to adjust the low CMRR_{el} by regulating the resistor R_8 and, consequently reduces the 50-Hz interference. This also becomes visible when we inspect the FFT (fast Fourier transform) of both signals as shown in Fig. 5.7:

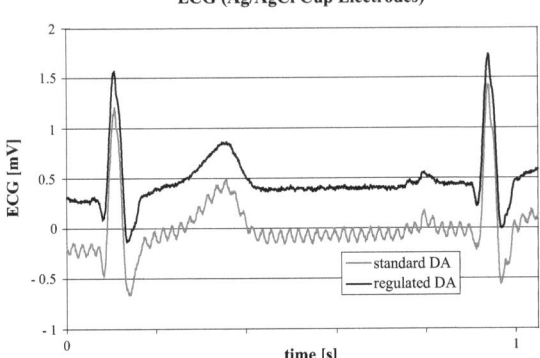

Fig. 5.6: An ECG recorded simultaneously with a conventional DA and the proposed regulated DA. The amplifying electrodes used have a different gain resulting in a low CMRR leading to power-line interference. An offset of +0.2 or −0.2 mV was added to the signals for better visualization.

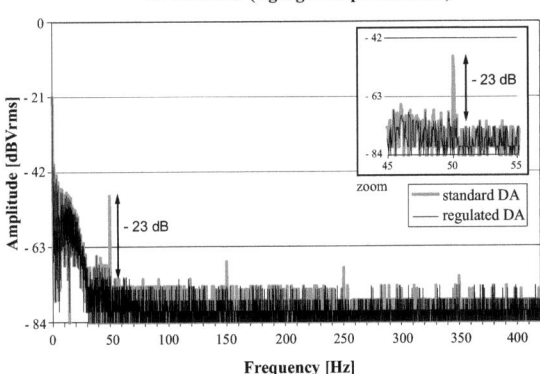

Fig. 5.7: The FFT of the signal using a regulated DA shows 23 dB less 50-Hz interference than the one using a conventional DA. The FFTs correspond to a signal segment of length 7 s sampled with 840 Hz.

The 50-Hz component is clearly reduced and the scalar value of the reduction is about -23 dB. It can also be seen that the frequency spectrum of the ECG signal is not altered by our method except that all odd harmonics of the 50-Hz power-line interference including the known 400-Hz common-mode signal V_{add} are removed. This demonstrates that our

5.3. Method

method is different from a simple analog 50-Hz notch filter.

Using the 50-Hz Interference as common-mode Signal

Again, the DA was designed on a test board using the same components as before. This time no common-mode voltage V_{add} was added. Instead, the 50-Hz interference was measured at two different locations as described in section 5.3. The recordings in Fig. 5.8 are measured after an initial settling time of several seconds. For better visualization, we again added an offset of $+0.2$ or -0.2 mV. The result is comparable to the result of the first method as exemplified in the respective FFT displayed in Fig. 5.9.

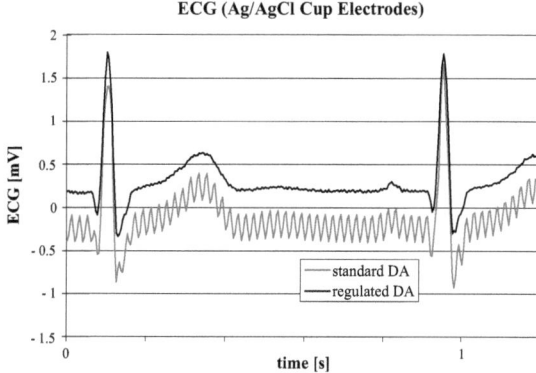

Fig. 5.8: An ECG recorded simultaneously with a conventional DA and the proposed regulated DA. The amplifying electrodes have a different gain resulting in a low CMRR leading to power-line interference. An offset of $+0.2$ or -0.2 mV was added to the signals for better visualization.

In this example our method reduces the 50-Hz power-line interference by about 30 dB. We can also see that the second harmonic (at 100Hz) is reduced, whereas all the other frequencies remain the same. The small differences at higher frequencies result from the noise level of the AD converters of the employed oscilloscope.

Comparing the Two Methods

The two methods (using a known common-mode signal versus using the power-line interference) yield a very similar result. The difference in

5. Gain Adaptation for Amplifying Electrodes

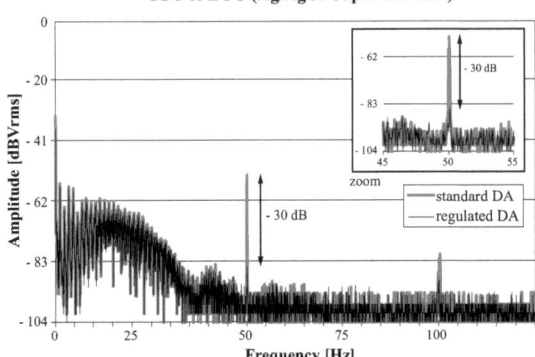

Fig. 5.9: The FFT of the signal after the regulated DA shows 30 dB less 50-Hz interference than that of the one recorded with a conventional DA. The FFTs correspond to a signal segment of length 24 s sampled with 250 Hz.

the FFT of the different recordings is mainly an offset due to the variation in amplitude and baseline of the two measurements. The reason for the difference in amplitude and baseline is that the measurements were taken on two different days and the exact location of the electrodes could not be reproduced. The question is why we need two different methods. The first method is much faster than the second. The frequency of the additional common-mode signal is eight times higher than the frequency of the power-line interference and its amplitude can be chosen to suit the specific situation. Consequently, the control loop settles in less than 100 ms, which corresponds to about 40 iterations. This is important in applications with many amplifying electrodes like that for an EEG [Mett 90].

Yet, the first method is based on the introduction of a known common-mode signal V_{add} applied to the input of the DRL circuit. This is not the case with the second method, which is simpler when looking at the analog part of the circuit. This method does not require the use of a DRL circuit, a simple reference electrode would be sufficient (although in our example we use a DRL circuit). On the other hand, the second method requires much more time for the control loop to settle. Due to the very low signal level and the ECG being in the same frequency band, one step of the PD controller occurs only after at least 20 cycles (400 ms). The control loop still needs about 40 iterations, which takes about 16 seconds. There is no prior knowledge required of the phase of the common-mode signal.

5.4 Novelty

We have successfully shown that the known principle of increasing the CMRR of a differential amplifier can also be applied to a two-stage amplifier. As the method does only change the gain of the second stage it is possible to place the first stage of the amplifier on the electrodes, thus having amplifying electrodes. The use of amplifying electrodes without a common reference point and without a possibility to compensate for their individual gains would result in an overall CMRR which is lowered.

It is important to note that our method also functions with electrodes which do not share a common reference node (as for example digitizing electrodes). As there is no need for a common reference node, the number of wires needed for the amplifying electrodes can be reduced by one as compared to the same design not using automated gain adaptation.

It is also the first method known to the author using the power-line interference itself as the input to generate a control signal. This allows us to use the method without the need of a DRL circuit (although the authors recommend a DRL circuit).

6. A PSEUDO-DIFFERENTIAL AMPLIFIER FOR BIOELECTRIC EVENTS WITH DC-OFFSET COMPENSATION USING TWO-WIRED AMPLIFYING ELECTRODES

This chapter is a synopsis of the corresponding publication in the IEEE Transactions on Biomedical Engineering [Dege 06].

6.1 Problem Statement

Amplifying electrodes supposedly reduce the input-referred noise of an amplifier for bioelectric events because the signal is amplified at the source already as opposed to standard amplifiers where the electrodes act as (passive) transducers for the bioelectric signals.

Yet, amplifying electrodes have several limitations. First, the gain of the electrodes is limited by the electrode-electrolyte voltage which can vary between electrodes by up to ± 300 mV [AAMI 99]. If this offset is not compensated for, active electrodes with a supply voltage of 3 V will have their gain limited to 14 dB. Second, the gain mismatch between electrodes will limit the CMRR as expressed in equation (2.23). For example, according to equation (2.27) using 1-% resistors to set the gain of non-coupled electrodes to 14 dB will result in an expected worst-case CMRR of 29.9 dB (dependent of the gain). Additionally, the use of amplifying electrodes will result in more wires and larger connectors. This may lead to stiffer cables and increased mechanical size of the connectors, especially in the case of EEG amplifiers with a large number of electrodes, e.g., 256 for a dense-array EEG. Stiffer cables in turn will lead to increased motion artifacts.

The following work demonstrates two-wired amplifying electrodes with a gain of 40 dB together with an amplifier for bioelectric events making use of these electrodes. The overall CMRR (@ 50 Hz) of the amplifier with DRL and gain adaptation is 109.7 dB.

6.2 Prior Art

Amplifying electrodes with a gain greater than 0 dB are rarely described in literature. An example of such an electrode is derived from the standard two op-amp instrumentation amplifier with the first stage being placed onto the electrodes [Valc 04]. The resulting amplifying electrode requires four wires for the reference electrode and five wires for the signal electrodes. The gain of the active electrode was 15 dB and the CMRR of the amplifier was 96 dB. The CMRR is kept high by keeping a common reference node for all electrodes [Pall 91b], which on the other hand leads to the requirement of five wires per signal electrode.

Another amplifier was described using amplifying electrodes with 20 dB gain and four wires (plus shield) but without a common reference node. The amplifier has a CMRR of 85-90 dB [Duns 95]. The relatively high CMRR was achieved by hand-picking the gain-setting resistors of the electrodes while manually trimming the gain-setting resistors of the instrumentation stage (this corresponds to a manual implementation of the automatic method we earlier described in chapter 5).

One publication describes amplifying electrodes with a highpass filter placed on the electrode itself. The electrodes have been built using small SMD components. These electrodes have a mid-band gain of 43.5 dB [Mett 97]. The CMRR of an amplifier using these electrodes is not given in the publication. The electrodes require four wires but do not allow temporary adjustment of the corner frequency of the highpass filter. (This is an often desired feature to reduce the recovery time after placement or after a motion artifact.)

Electrodes with two wires but only 0 dB gain are described in literature [Padm 90], [Dege 07]. An implementation of such two-wired buffer electrodes are even commercially available [bios].

6.3 Method

Having a large gain on the electrodes would be beneficial for three reasons. First, using the same op-amp (operational amplifier) with a higher gain reduces the input-referred noise of the op-amp itself as confirmed by noise measurements (see 6.3.3 Fig. 6.8). Second, the noise contribution of the subsequent stages to the input-referred noise is reduced the more the higher the implemented gain in the first stage is. The noise requirements of the following amplifier stage can be lowered accordingly, this might also allow for a reduction in power consumption. Third, compared to using buffer electrodes the total number of op-amps can be reduced because a large part of the gain is implemented on the electrodes.

To reduce the number of wires, we will use a current supply instead of a voltage supply [Padm 90], [Dege 07].

To compensate for any gain mismatch between the amplifying electrodes, we will measure the latter and compensate for the difference in gain later, using a software algorithm. This is only possible because the amplifier developed for the electrodes has a pseudo-differential topology. The input signals are sampled using differential AD converters. This allows placing the differentiation between two signals into the software part of the recording system. The method used for compensating a mismatch in gain is comparable to the hardware-based method described in chapter 5.

The compensation of the electrode-electrolyte offset voltage is achieved by a DC compensation based on an error-feedback loop which conserves the high input impedance of the amplifying electrodes (no highpass filter at the input of the amplifier). At the same time, the error signal represents the low-frequency part of the bioelectric signal (frequency below 0.016 Hz for ECG respectively below 1 Hz for EEG) which may be clinically relevant and is normally lost when using a highpass filter at the input of an amplifier. The low-frequency part of the bioelectric signal can therefore be measured down to DC, although at a lower resolution. (For examples of relevant data refer to section 2.3.5.)

6.3.1 Two-Wired Amplifying Electrodes

The amplifying electrodes are built using a CMOS-input op-amp, i.e., the LMC7111 (National Semiconductor). The LMC7111 was chosen because of the low supply voltage of 2.7 V and the high input resistance. The op-amp is used in a non-inverting configuration with the gain set by two resistors (1%) to 40 dB. A diode may be used to ensure proper start-up behavior. The full circuit is depicted in Fig. 6.1.

Instead of a voltage supply the electrode receives a constant current I_{sup}. The current is then split between the different parts of the circuit. I_L denotes the supply current of the op-amp (LMC7111), I_D the current through the diode D_1 and I_R the current through the gain-setting resistive network built by R_1 and R_2.

The role of the diode is mainly to insure a stable start-up behavior and may or may not be necessary depending on the topology of the employed op-amp. We tested each employed op-amp with or without the diode D_1 before deciding on the topology. We will give an overview of the tested op-amps in table 6.2.

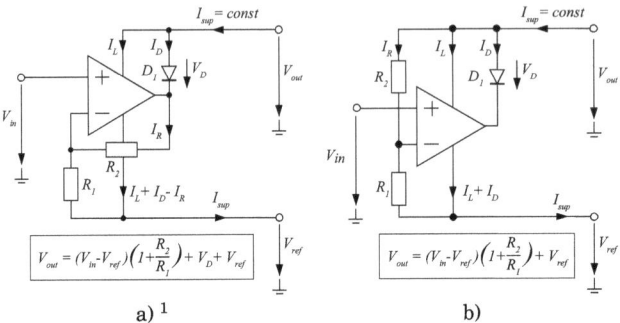

Fig. 6.1: The principle of the two-wired amplifying electrode biased by a constant current-source I_{sup}. The active part of the electrode consists of a CMOS op-amp, two resistors to define the gain and a diode used for improved start-up behavior. Part a) depicts the published version which has an additional voltage offset of V_D. Part b) shows the improved design with lower supply voltage. The mass symbol refers to a common reference point, e.g., the negative pole of the battery.

The negative supply is connected to a reference voltage V_{ref} which is typically held around 500 mV above the system ground and will play an important role in the DC-offset compensation scheme.

All measurements presented herein are made with the electrode depicted in Fig. 6.1 a). For a new design we strongly recommend using the improved version depicted in Fig. 6.1 b). All calculations in this chapter are made for the improved version of the electrode.

The output voltage of the electrode is given by:

$$V_{out} = (V_{in} - V_{ref})\left(1 + \frac{R_2}{R_1}\right) + V_{ref} \qquad (6.1)$$

This equation is true as long as the current-source delivers enough current to supply the active electrode but not more than the op-amp can sink and as long as the open-loop gain of the op-amp is substantially higher (> 40 dB) than the gain of the electrode. :

$$I_{sup} > \frac{V_{sup\ max}}{R_1 + R_2} + I_{L\ max} \qquad (6.2)$$

$$I_{sup} < \frac{V_{sup\ min}}{R_1 + R_2} + I_{sink\ max} \qquad (6.3)$$

The amount of current an op-amp can sink may depend on the supply voltage V_{sup} of the op-amp. The supply voltage of the op-amp on the

6.3. Method

other hand depends on the input signal:

$$V_{sup} = V_{out} - V_{ref} = (V_{in} - V_{ref})\left(1 + \frac{R_2}{R_1}\right) \quad (6.4)$$

V_{sup} is the supply voltage of the op-amp which has both an upper and lower limit, which can be found in the corresponding data sheet. The supply voltage of the op-amp has always to be above the minimal supply voltage specified in the data sheet. The upper limit is either given by the data-sheet or is limited by the supply voltage of the whole circuit.

$$V_{out\ min} - V_{ref\ max} \geq V_{sup\ min} \quad (6.5)$$

Where $V_{sup\ min}$ denotes the minimal supply voltage of the op-amp. The value depends on the amount of supply current I_{sup} the op-amp has to sink (see Fig. 6.7). As an example for the LMC7111 the minimal supply voltage was measured to be 2.6 V when sinking a current I_{sup} of 1 mA.

To ensure the correct behavior of the amplifying electrode, we suggest to test the electrode with half the supply current at the maximal output voltage and with the double supply current at the minimal output voltage (similar to a corner analysis).

The output resistance R_{out} of the active electrode can be expressed by the output resistance of the op-amp $R_{out\ Op}$ plus the output resistance of the diode $R_{out\ D}$ (if required) divided by the open loop gain A_0 of the op-amp:

$$R_{out} = \frac{R_{out\ Op} + R_{out\ D}}{1 + A_0} \quad (6.6)$$

As an example, the open-loop gain of the LMC7111 for $V_{sup\ OP} = 2.7$ V while sinking a current is typically $> 100\ dB$. $R_{out\ Op}$ is normally about 100 Ω.

The differential resistance of the diode on the other hand depends on the current through the diode I_D and can be approximated by:

$$R_{out\ D} = \frac{\partial V_D}{\partial I_D} = \frac{\partial \left(V_{th} \ln\left(\frac{I_D}{I_S}\right)\right)}{\partial I_D} = \frac{V_{th}}{I_D} \quad (6.7)$$

$V_{th} = k_B T/q_0$ is the thermal voltage and yields about 25 mV for an ambient temperature of 22° C (295.15° K). I_S is the reverse current of the diode. We used a small-signal diode BAS16 (Fairchild Semiconductor) which has a reverse current of about 1 μA

For our electrode with a supply current I_{sup} of 1 mA, we measured an output resistance of the diode of about 25.6 Ω which is very close to the 25.4 Ω resulting from equation (6.7).

Going back to equation (6.6), we can estimate the total output resistance of the amplifying electrode to be far below 0.1 Ω.

The amplifying electrodes were supplied by a current-source of the type J505 (Vishay) which was already used for the buffer electrodes in chapter 4. Again, the current-source (J-FET) can be considered to be the upper part of the output stage of the op-amp. The requirements for the op-amp are that one supply is included in the input range and that the maximal current the op-amp output stage can sink exceeds the value of the supply current I_{sup} as defined in equation (6.3).

6.3.2 Amplifier Stage

The amplifier stage uses one DRL electrode, one reference electrode (subscript 'r') and 'n' signal electrodes to measure the bioelectric signal of interest (see also section 2.3.3). For the calculations we consider one signal electrode out of the 'n' electrodes (subscript 'i'). The bioelectric signal of interest is then $V_{in\,i} - V_{in\,r}$. Fig. 6.2 depicts the amplifier stage.

The primary requirement of the amplifier is to ensure that the output voltages of the electrodes do not saturate. This is a stumbling block for the use of amplifying electrodes with a gain larger than 15 dB (depending on the supply voltage). To avoid saturation of the output voltage we will adapt the reference voltage $V_{ref\,i}$ of each individual signal electrode 'i' which is simultaneously the lower (negative) supply voltage of the op-amp placed on the amplifying electrode. We can condense the function of the amplifier stage into three main tasks:

- Provide the necessary supply current I_{sup} for the electrodes.

- Generate the adequate reference voltages $V_{ref\,i}$ in order to compensate for any DC offset between electrodes.

- Reduce any common-mode voltage by feeding it back to the body via a DRL circuit which is simultaneously determining the DC component of the input voltage of the reference electrode.

These three functions are achieved in the above order by a current-source and two different feedback loops.

The current-source for each electrode is a J505 (Vishay), based on a single J-FET (Junction FET). The J505 was chosen for its simple topology. To source 1 mA the J505 requires a voltage drop of about 1.2 V. (For the schematic of the current-source please refer to Fig. 4.5).

6.3. Method

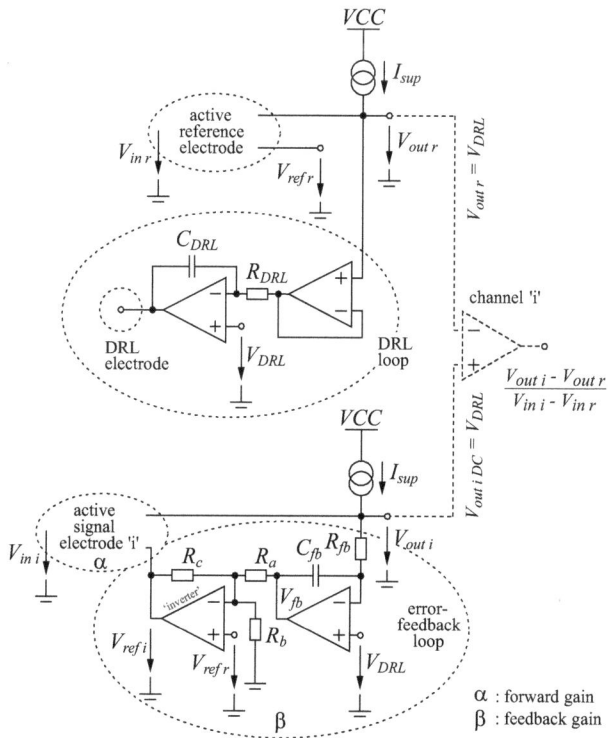

Fig. 6.2: The amplifier stage for providing the necessary bias voltages and bias currents. Bias sources are implemented by a current-source and two different feedback loops. The first feedback loop ('DRL loop') is built around the reference electrode and the DRL electrode. The second loop ('error-feedback loop') is built around each signal electrode of which only the electrode 'i' is drawn. The amplified output signal for each channel is $V_{out\,i} - V_{out\,r}$. All op-amps are of the LT1490 type. The ground symbol stands for a local reference, e.g., the system ground of the amplifier.

The first feedback loop ('DRL loop') is realized around a reference electrode and the DRL electrode. The purpose of the first loop is to define the correct bias voltage for the reference electrode. In detail this means that if the output voltage $V_{out\,r}$ of the reference electrode is higher than V_{DRL}, a current will flow into the resistor R_{DRL} and the op-amp which drives the DRL electrode has to lower its output voltage to allow this current to flow through C_{DRL} as well. This lowers the input voltage of all electrodes connected to the DRL electrode via the human body. Yet lowering the input voltage of the reference electrode $V_{in\,r}$ while the reference voltage $V_{ref\,r}$ stays fixed at 500 mV will also lower the output voltage V_{out} and close the loop. The value of V_{DRL} was chosen in such a way that it was in the middle of the valid range of $V_{out\,r}$ (this will be analyzed further in section 6.3.3 where the results are described). For the amplifying electrode in Fig. 6.1 a) V_{DRL} should be about 3.55 V, for the amplifying electrode in Fig. 6.1 b) V_{DRL} should be about 3 V.

In other words, the task of the first feedback loop is to maintain the voltage between the system ground of the amplifier and the surface potential of the body at a constant level, and simultaneously reduce any common-mode signal at the input. The input voltage of the reference electrode $V_{in\,r}$ will also be fixed by this first feedback loop. For the present circuit $V_{in\,r}$ is held around 25 mV above V_{ref}, the exact value will be derived later in equation (6.1).

The corner frequency of the first feedback loop is $f_0 = 1/2\pi R_{DRL} C_{DRL}$. To guarantee the stability of the first loop we chose the values $R_{DRL} = 220$ kΩ and $C_{DRL} = 15$ nF, which resulted in a corner frequency of $f_0 = 53$ Hz, an open-loop phase margin of about 30° [Wint 83] and a CMRR improvement of about 40 dB (@ 50 Hz) as indicated by equation (2.46) and visible in Fig. 6.9 later in the results section. Smaller values of C_{DRL} will increase the overall CMRR but may lead to oscillation or instability within the feedback loop.

The second loop ('error-feedback loop') is built individually around each signal electrode. The purpose of the second loop is to compensate for a difference in the electrode-electrolyte interface voltage between the reference electrode and each signal electrode. The voltage difference may reach values of up to ± 300 mV [AAMI 99].

In other words, the second loop implements a highpass behavior for each signal electrode. We will examine the loop in detail for the ith electrode. The reference voltage $V_{ref\,i}$ is adjusted by the loop as long as the DC value of the output voltage $V_{out\,i}$ is different from V_{DRL}. The guiding principle is the same as for the first feedback loop except that this time it is the reference voltage which is adapted while the input voltage $V_{in\,i}$ is fixed (by the first feedback loop superimposed by the bioelectric signal). For the loop to be stable we need to invert the output of the corresponding integrator. For this purpose we added a voltage inverter with the following transfer function:

$$V_{ref\,i} = \frac{R_c + R_b}{R_b} V_{ref\,r} - \frac{R_c}{R_a}(V_{fb} - V_{ref\,r}) \tag{6.8}$$

The inverter will also convert the range of V_{fb} which is 0 to VCC to the desired reference voltage swing $V_{ref\,i} = 500 \pm 300$ mV by its gain

6.3. Method

of $-R_c/R_a$. As we will see later in equation (6.11), this gain will also reduce the size of the feedback capacitor for a chosen time constant.

By choosing an adequate value for R_b, the midpoint of the output range can be shifted to $V_{ref\ r}$. As a consequence the initial reference voltage $V_{ref\ i}$ after power-up will be close to $V_{ref\ r}$, thus reducing the start-up time of the circuit.

The output-range limitation of the inverter will also ensure that $V_{ref\ i}$ cannot be substantially higher than V_{DRL}. This is important because if $V_{ref\ i} - V_{DRL} \geq 550$ mV, the built-in protection diode of the op-amp placed on the reference electrode would start to conduct a current and thus prevent the operation of the first feedback loop.

The values used were $R_a = 47$ kΩ, $R_b = 9.1$ kΩ and $R_c = 4.7$ kΩ. The second loop forces the DC value of $V_{out\ i}$ to be equal to V_{DRL} plus a residual offset. The offset is mainly given by the input offset and bias current of the op-amp used and can be expressed by:

$$V_{out\ i\ DC} = V_{DRL} + V_{off\ opamp} + R_{fb}\ I_{bias\ opamp} \qquad (6.9)$$

$V_{off\ opamp}$ is the input offset voltage of the op-amp building the integrator, $I_{bias\ opamp}$ corresponds to the bias current at the negative input. In our prototype circuit the value of the residual offset was about 40 mV.

The time constant of the second feedback loop is much lower (at least five decades) than the time constant of the first feedback loop. This ensures the stability of the second feedback loop despite it being interlinked with the first one (see also section 6.3.2, Fig. 6.9).

The amplifier stage is of a pseudo-differential nature, the amplified and DC-offset-free bioelectric signal is represented by $V_{out\ i} - V_{out\ r}$. Yet the bioelectric signal is only visible on $V_{out\ i}$ whereas the common-mode signal is present on both $V_{out\ i}$ and $V_{out\ r}$. Considering only $V_{out\ i}$ would show large common mode interferences (see also Fig. 6.10).

Fig. 6.3 illustrates the overall transfer function of the pseudo-differential amplifier and the electrodes.

The transfer function can be expressed by the forward gain α and the feedback gain β:

$$\frac{V_{out\ i} - V_{out\ r}}{V_{in\ i} - V_{in\ r}} = \frac{\alpha}{1-\alpha\beta} \qquad (6.10)$$

Where α corresponds to the gain of the electrodes as stated in equation (6.1). The bandwidth $f_{0\ lp}$ of α is given by the GBP (gain-bandwidth-product) of the employed op-amp (LMC7111) which is 50 kHz divided by the gain of the amplifying electrodes. β corresponds to the transfer

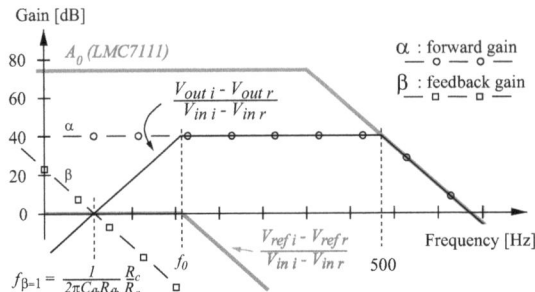

Fig. 6.3: The transfer function of the pseudo-differential amplifier stage with the electrodes. The upper corner frequency $f_{0\,lp}$ is fixed by the function of the open-loop gain A_0 and the corresponding gain-bandwidth-product (GBP) of the employed op-amp (LMC7111). The overall gain is expressed by equation (6.10) and the lower corner frequency f_0 depends on the feedback loop as expressed in equation (6.11).

function of the integrator multiplied by a constant factor R_a/R_c as expressed by equation (6.8). The resulting transfer function is a highpass where the corner frequency is defined by the condition $\alpha\beta = 1$. Consequently, we can express the cut-off frequency f_0 by:

$$f_0 = \frac{1}{2\pi R_{fb} C_{fb}} \left(1 + \frac{R_2}{R_1}\right) \frac{R_c}{R_a} \qquad (6.11)$$

Below the corner frequency the transfer function can be approximated by $1/\beta$ and above f_0 by α. We used $R_{fb} = 10$ MΩ and $C_{fb} = 10$ μF which results in a corner frequency of $f_b = 0.016$ Hz as required for recording an ECG. To restore the DC value after replacing the electrodes or to allow for fast recovery after an artifact, we placed a switch in parallel to R_{fb} enabling us to temporarily shift the corner frequency. When opening the switch a small charge may be injected into C_{fb} resulting in a voltage step at the output. This has no adverse effects on the system.

A particularity of this amplifier is the fact that the cut-off frequency of the highpass depends on the gain of the active electrode. We can use the same amplifier for EEG by using a low-noise electrode with a gain of 60 dB which results in a cut-off frequency of 0.16 Hz, which is lower than most EEG amplifiers offer.

DC-Offset Measurement

As reported earlier, the purpose of the second feedback loop is to adjust for the electrode-electrolyte voltage offset between the reference elec-

6.3. Method

trode and each individual signal electrode. Let us assume that there is a DC offset between $V_{in\ i}$ of the ith signal electrode and $V_{in\ r}$ of the reference electrode. The second feedback loop will then adjust the reference voltage $V_{ref\ i}$ until the DC component of the signal is again V_{DRL}. This leads to the following relation:

$$V_{ref\ i} - V_{ref\ r} = \left(1 + \frac{R_1}{R_2}\right)(V_{in\ i} - V_{in\ r}) \qquad (6.12)$$

$$= \left(\frac{\text{gain}}{\text{gain} - 1}\right)(V_{in\ i} - V_{in\ r}) \qquad (6.13)$$

For a large gain $V_{ref\ i} - V_{ref\ r}$ equals $V_{in\ i} - V_{in\ r}$, which corresponds to the offset voltage between electrodes. This means that by measuring the reference voltage of an electrode, it is possible to deduce the offset voltage of this electrode compared to the reference electrode. This feature allows measuring the full bandwidth of the bioelectric signal (plus the electrode-electrolyte offset voltage, which is considered to be relatively stable). Fig. 6.3 also displays the frequency response of equation (6.13) which can is expressed by:

$$\frac{V_{ref\ i} - V_{ref\ r}}{V_{in\ i} - V_{in\ r}} = \frac{\alpha\beta}{1 - \alpha\beta} \qquad (6.14)$$

V_{fb} holds the same information but with an amplitude ten times larger. To our knowledge this is the first circuit reported which allows us to measure a bioelectric signal with frequencies down to DC while suppressing the DC offset between electrodes.

A second consequence of equation (6.13) is that the limits within which the circuit can compensate a DC offset are given directly by the voltage swing of $V_{ref\ i}$ which on the other hand is limited by the total supply voltage divided by the constant factor R_c/R_a as shown in equation (6.8).

Amplifier Design and CMRR

The main problem of amplifying electrodes is the possible reduction of the CMRR due to a variance in gain (@ 50 Hz) between electrodes. Let us assume that the reference electrode has a gain of G_r and the signal electrode 'i' has a gain of G_i. If we now calculate the difference we obtain:

$$V_{out} = G_i V_{in\ i} - G_r V_{in\ r} \qquad (6.15)$$

this can be expressed as:

$$V_{out} = \frac{G_i + G_r}{2}(V_{in\ i} - V_{in\ r}) + (G_i - G_r)\frac{V_{in\ i} + V_{in\ r}}{2} \qquad (6.16)$$

This is nothing else than separating the differential-mode gain G_{DM} and common-mode gain G_{DM} knowing that the differential-mode and the common-mode input voltages are given by:

$$V_{DM} = V_{in\,i} - V_{in\,r} \qquad (6.17)$$

$$V_{CM} = \frac{V_{in\,i} + V_{in\,r}}{2} \qquad (6.18)$$

Thus equation (6.16) leads directly to the CMRR:

$$\text{CMRR} = \frac{\text{Gain}_{DM}}{\text{Gain}_{CM}} = 20\log\left(\frac{G_i + G_r}{2\,|G_i - G_r|}\right) \qquad (6.19)$$

Which corresponds to the formula derived in section 2.2.4. It allows us to calculate the expected CMRR for a given gain mismatch [Pall 91]. If the gain mismatch is a result of the tolerance of the gain-setting resistors we can express the worst-case CMRR of the total INA by equation (2.38). For our circuit we use resistors with a tolerance of 1% and electrodes with a gain of 40 dB which leads to an expected worst-case CMRR of about 28 dB of the total INA due to the use of the amplifying electrodes (which are the limiting factor). Note, our electrodes are non-coupled electrodes, using coupled electrodes would yield a better CMRR.

28 dB may be sufficient for a body-worn device when using a DRL circuit but not for a clinical ECG or EEG (see table 2.2 for an overview). A DRL circuit may improve this CMRR by up to 50 dB [Mett 90]. The DRL circuit implemented for this amplifier results in an improvement of about 40 dB (@ 50 Hz) pushing the CMRR of the amplifier (including DRL) to about 68 dB (@ 50 Hz).

To improve the CMRR further we will compensate for the gain mismatch of each signal electrode in respect to the reference electrode when differentiating, similar to the method presented in chapter 5.

Because our amplifier is pseudo differential (the signal is only present of $V_{out\,i}$ but the common mode interference also present on $V_{out\,r}$) we can first convert the signals into a digital representation and then still compensate for the gain mismatch in software. Therefore we require all signals to be sampled, either by a fully single-ended ADC (analog to digital converter) and a multiplexer or in a pseudo-differential configuration as shown in Fig. 6.4. (Using two fully differential converters is also possible but would no longer allow compensating for a gain mismatch in software.)

The pseudo-differential approach requires one more AD channel for the reference V_{DRL}. The reference voltages $V_{ref\,i}$ and $V_{ref\,r}$ are sampled only if the low-frequency part of the bioelectric signal (below f_0) is of interest.

6.3. Method

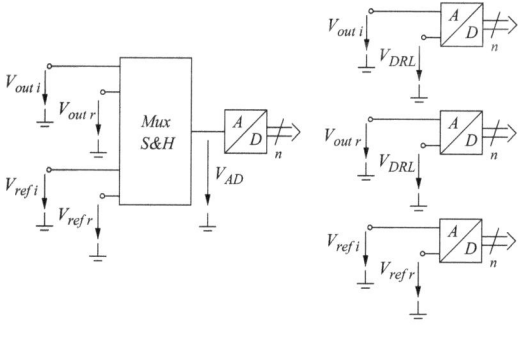

a) fully single ended   b) pseudo differential

Fig. 6.4: Two ways of interfacing the amplifier by a) converting all four single-ended signals separately or b) using three differential AD converters to convert them in a pseudo-differential way. This enables simple software based algorithms to compensate for differences in gain and thus improving the CMRR.

To compensate for the gain mismatch (@ 50 Hz) we will first measure the amplitude of the power-line interference $A_{50\,r}$ and $A_{50\,i}$ for the different signals. The amplified and compensated bioelectric signal V_{BE} is then calculated by:

$$V_{BE} = 2\,\frac{A_{50\,r}V_{out\,i} - A_{50\,i}V_{out\,r}}{A_{50\,r} + A_{50\,i}} \qquad (6.20)$$

$$= 2\,\frac{G_i A_{50\,r}V_{in\,i} - G_r A_{50\,i}V_{in\,r}}{A_{50\,r} + A_{50\,i}}$$

$$= 2\,\frac{G_i G_r A_{50}(V_{in\,i} - V_{in\,r})}{A_{50}(G_r + G_i)}$$

$$= 2\,\frac{G_i G_r (V_{in\,i} - V_{in\,r})}{G_r + G_i} \qquad (6.21)$$

G_r and G_i represent the gain (with mismatch) of the two amplifying electrodes considered, one being the reference electrode the other the ith signal electrode. $A_{50\,r}$ and $A_{50\,i}$ stand for the measured amplitude (by FFT) of the 50-Hz power-line interference overlaid on the output voltage of the reference electrode, respectively the ith signal electrode. A_{50} would then be the amplitude of the 50-Hz common mode interference on the body prior to the amplification. Note, A_{50} is not present in the final equation (6.21).

To obtain $A_{50\,r}$ and $A_{50\,i}$ we used the discrete Fourier transform of the sampled values. Unlike the method described in chapter 5 this method does not necessitate a phase-shift measurement because the Fourier transform allows us to obtain the amplitude of the 50-Hz component independently from any phase shift (we assume that the 50 Hz are well within the passband of the amplifier).

By implementing equation (6.21) based on the measurement of the 50 Hz interference, the latter can be reduced as shown in section 6.3.3. The remaining 50-Hz component of the output signal is mostly part of the differential bioelectric signal itself. An example is shown in Fig. 6.12 and Fig. 6.13.

The CMRR is improved to the value of:

$$\text{CMRR} = \frac{G_i A_{50\,r} + G_r A_{50\,i}}{2|G_i A_{50\,r} - G_r A_{50\,i}|} \qquad (6.22)$$

An absolute value for the improved CMRR cannot be given, because the improvement also depends on the value of the measured common-mode signal itself. The larger the common-mode signal, the better the CMRR. We can say that the CMRR increases as much as necessary to ensure that the measured common-mode signal is minimized on the reading.

We can give an upper limit to which this schema can function. The maximum value for the CMRR improvement is obtained if the common-mode signals measured on the two output voltages $A_{50\,i}$ and $A_{50\,r}$ are as large as the input range of the ADC. The 50-Hz interference after the gain adaptation will correspond to the resolution of the ADC, and, as a result, the upper limit can be given by:

$$\text{CMRR} \leq 20 \log\left(2^n\right) \qquad (6.23)$$

Where 'n' corresponds to the resolution of the ADC. It has to be noted that this is not necessarily the best measurement situation, it is more advantageous to have less power-line interference from the beginning even though this yields a lower CMRR.

Contrary to a 50-Hz notch filter, the method will reduce all common-mode signals without altering the differential-mode signal, i.e., the bio-electric signal.

The difference between the fully single-ended circuit shown in Fig. 6.4 a) and the pseudo-differential implementation of Fig. 6.4 b) presented in Fig. 6.4 is the necessary input range of the ADC. In Fig. 6.4 a) the input range must be at least $V_{DRL} + V_{out}$ whereas in Fig. 6.4 b) the input range can be as low as $\pm V_{out}/2$. As a consequence, the resolution for the same number of bits is higher in b). For example, the AD7731 can have an input range as low as ± 20 mV and provides three differential ADCs in one IC. At 400 Hz the AD7731 has a resolution of 14.5 bit which is equivalent to 1.73 μV. The upper limit of the CMRR improvement according to equation (6.23) would then amount to 87 dB (in addition to

6.3. Method

the expected 68 dB CMRR of the amplifier with DRL mentioned above @ 50 Hz).

Digital Control Loop

Regardless of which AD-conversion scheme is chosen, it is possible to implement the second feedback loop responsible for the DC-offset compensation by an analog circuit as shown in Fig. 6.2 or by a digital filter as shown in Fig. 6.5.

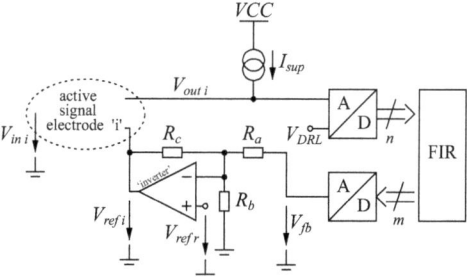

Fig. 6.5: The second feedback loop built as a digital control loop by replacing the integrator of Fig. 6.2 with a digital filter (FIR) and an AD/DA converter. The feedback loop is again shown for the electrode 'i'. There is *no* need for a capacitor.

The partially digital loop functions in a similar way to the fully analog loop. The feedback voltage V_{fb} is generated by the DAC (digital to analog converter) while the digital filter (FIR) generates the necessary codes for the DAC based on the differential input of the ADC. The second op-amp stage (inverter) is still needed for sinking the supply current of the amplifying electrode and serves also for adapting the output range of the DAC to the desired reference voltage range. Equation (6.8) is still valid and defines the function of the inverter.

The corner frequency of the digital filter $f_{0\ FIR}$ is calculated analogously to the corner frequency of the integrator used in the analog loop. Starting from equation (6.11) we then obtain:

$$f_{0\ FIR} = f_0 \frac{R_1}{R_1 + R_2} \frac{R_a}{R_c} \qquad (6.24)$$

f_0 stands for the desired corner frequency of the transfer function of the amplifier (e.g., 0.016 Hz for ECG). R_1 and R_2 are the gain-setting resistors of the amplifying electrode.

The main differences between the analog and the digital implementation are:

- the digital loop requires no capacitor, which enables fully integrated solutions, e.g., within an ASIC.

- the time constant between different channels matches perfectly and removes the CMRR degradation due the mismatch of corner frequencies (see also section 2.2.4)

- the digital loop introduces discrete steps in the output waveform (quantization noise). In between these steps the signal does not show the typical superimposed exponential variation which is usually a characteristic of highpass filtered signals. Because the exponential variation is absent it is possible to measure signal features like the S-T elevation without distortion. Saturation of the output voltage can be avoided easily.

In addition, adaptive filter techniques may be used resulting in shorter settling times. Likewise the part count decreases as the DAC replaces an op-amp together with some passive components as well as the ADC necessary to sample V_{fb} (required only if the DC part of the signal is of interest).

Because of the discrete nature of the output voltage V_{fb}, the DC offset of the bioelectric signal cannot be completely nullified. There remains a residual offset $V_{off\ res}$ which can be calculated in relation to the resolution of the DAC:

$$V_{off\ res} = \frac{\text{VCC}}{2^m}\left(1 + \frac{R_2}{R_1}\right) \cdot \frac{R_a}{R_c} \qquad (6.25)$$

m denotes the number of bits of the DAC. For a supply voltage of 5 V and a DAC with 12 bits this results in a residual offset voltage of ±12 mV. In other words, each time the digital code of the DAC is changed by the FIR filter, the signal $V_{out\ i}$ will display a discrete step (quantization noise) with an amplitude corresponding to a multiple of the residual offset calculated in equation (6.25).

If not handled appropriately, this leads to an oscillating behavior of the output voltage $V_{out\ i}$ with a mean value of zero volt. This behavior is shown in section 6.3.3 Fig. 6.14. Because the exact timing of these steps is known a priori by the software, the latter can be removed before displaying or storing the signal. A better solution is to implement a hysteresis where the two thresholds differ at least by the value of the residual offset calculated in equation (6.25).

6.3.3 Results

Two-Wired Amplifying Electrodes

The fabrication of the prototype electrode was straight-forward. We mounted all the components on a printed circuit board (PCB). The PCB was then soldered onto the back of a snap button cut from a pair of trousers which allowed us to use different kinds of disposable electrodes. An example of a disposable pre-gelled EEG electrode clipped to an active electrode is shown in Fig. 6.6. One wire is connected to a low-ohmic node (i.e., the reference voltage) and was used to shield the output voltage of the amplifying electrode. This yields a relatively large input capacitance for the subsequent stage, in this case an ADC, lowering the input impedance of the second stage. This has no negative effect because of the low output resistance of the amplifying electrode.

Fig. 6.6: A shielded amplifying two-wired active electrode mounted onto a button in order to attach different types of disposable electrodes.

First we measured the input range for which the electrode operates in the linear region. Fig. 6.7 shows the output voltage V_{out} while varying the input voltage V_{in}, the reference voltage V_{ref} was set to zero.

The lower corner of the input voltage range is defined by the minimal supply voltage of the employed op-amp which is required to sink the current present at the output, i.e., $I_{sup} - I_L - I_R$. It is important to note that for both op-amps the required supply voltage increases several hundred millivolts if the supply current increases from 1 mA to 2 mA.

The upper corner of the input voltage is defined by the available voltage supply. In our case we had a voltage supply of 5 V where 1.2 V were dropped over the current-source (i.e., a J-FET J505 from Vishay). For a maximal output voltage of 3.8 V and a reference current of 1 mA, the linear output range was about 1.25 V starting from 2.5 V. This translates to an input range of 12.5 mV starting at 20 mV for the design a) or at 25 mV for the design b). The input range of the design a) starts from

6. Two-wired Amplifying Electrodes

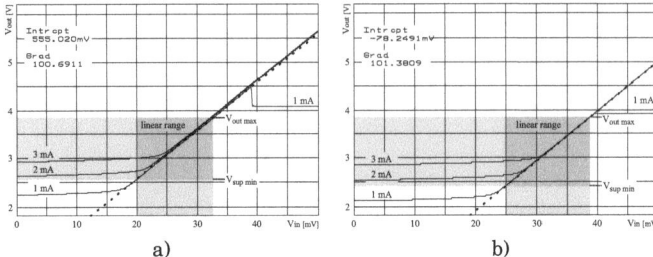

Fig. 6.7: The characteristic voltage gain of the two-wired amplifying electrode measured for three different supply currents and $V_{ref} = 0$. Part a) depicts the published electrode from Fig. 6.1 a) built with an op-amp LMC7111 (National Semiconductor) and a diode which is a BAS16. Part b) shows the improved design from Fig. 6.1 b) built with an op-amp LMV301 (National Semiconductor) but without a diode. For both implementations R_1 and R_2 have a value of 39 Ω respectively 3.9 kΩ.

a lower voltage because of the additional voltage drop over the diode.

This input range may appear to be too small to accommodate the power-line interference. This is not the case: the DRL-loop will reduce the common-mode voltage between the body and the reference of the amplifier. In section 6.3.2 Fig. 6.9 we will see that the common-mode gain of the amplifier is approximately one (@ 50 Hz). This means that common-mode voltage seen at the input is hardly amplified by the different electrodes. The common mode input range (@ 50 Hz) is therefore limited by the output swing of the DRL electrode, which is ± 1.5 V.

The gain of the amplifying electrode was set with two resistors of 39 Ω and 3.9 kΩ. The very small value of the resistors was chosen to minimize the voltage-noise contribution of the resistors in order to be able to measure the voltage noise of the op-amp. According to equation (2.1), the input-referred voltage-noise contribution of the two resistors in the frequency band from 0.1 Hz to 1 kHz amounts to about 25.4 nV each (the voltage noise of the larger resistors is divided by the gain of the amplifying electrode). For the chosen values the resistors will also define the minimal supply current $I_{sup\ min}$ which is defined by:

$$I_{sup\ min} \geq I_L + I_R = I_L(V_{sup\ max}) + \frac{V_{sup\ max}}{R_1 + R_2} \qquad (6.26)$$

For our example we have $I_L \approx 60$ μA (for the LMC7111) respectively $I_L \approx 150$ μA (for the LMV301) and $I_R \approx 840$ μA, which results in a minimal supply current $I_{sup\ min}$ of roughly 1 mA. This is certainly not optimal for a low-power application. For future designs the values of the resistors R_1 and R_2 should be higher, e.g., 1 kΩ and 100 kΩ.

The gain of the amplifying electrode appears in Fig. 6.7 as the gradient of the regression line. The gradient for a supply current of 1 mA is calculated by the parameter analyzer (HP4155B/4156B Hewlett Packard) and corresponds for this particular electrodes to a gain of 40.06 dB (LMC7111) and 40.12 dB (LMV301). The regression line in Fig.6.7 a) also reveals the voltage drop over the diode V_D which corresponds to the intersection with the y-axis and yields 555 mV.

For the supply current of 1 mA there is a sharp drop at $V_{out} \approx 4.5$ V. This is because when the current through the gain-setting resistors reaches about 1 mA the voltage of the negative input of the op-amp cannot rise any higher. In result the op-amp will short-circuit the diode in the attempt to increase the output voltage. The curve then becomes flat and stays at the value $V_{out} = (I_{sup} - I_L)(R_1 + R_2)$. This is of no consequence for our circuit as the voltage supply is limited by the current-source to 3.8 V (including the reference voltage V_{ref}.

A similar measurement (input current versus input voltage) was performed to determine the input resistance and the bias current of the electrode. The differential input resistance is over 1 TΩ and the bias current corresponds to about 60 pA.

It should be noted that for the improved amplifying electrode depicted in Fig. 6.1 b) the regression line ideally would not show any offset.

Noise

Next, we measured the input-referred spectral voltage-noise density of the amplifying electrode. The measurement is shown in Fig. 6.8.

The noise of the amplifying electrode is mainly due to the op-amp used, which in this implementation was an LMC7111 (the resistors R_1 and R_2 have been chosen accordingly). The total input-referred voltage noise of the electrode corresponds to the integrated spectral voltage-noise density over the desired bandwidth. For comparison, we used the extrapolated spectral voltage-noise density as indicated by the dotted lines. For a bandwidth of 10 Hz to 1 kHz (EEG) the computed integral results in 9.3 μV_{rms}. If we consider the total amplifier then the noise is mainly due to two sources: the active electrode and the noise generated by the error-feedback loop. This is because the noise of the error-feedback loop adds to the reference voltage $V_{ref\ i}$ and is therefore amplified by the active electrode. The total noise of the system in the bandwidth of 10 Hz to 1 kHz computes to 14.2 μV_{rms}. This is low enough for our test application which is a body-worn ECG recorder.

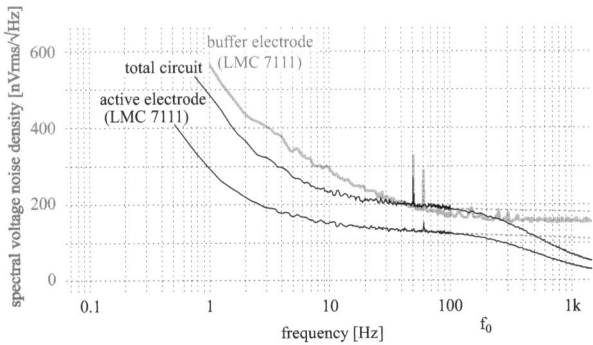

Fig. 6.8: The spectral voltage-noise density of the amplifying two-wired electrode and the total system (no additional gain) was measured with a vector signal analyzer (Stanford SR785). For comparison we added the spectral voltage-noise density of the two-wired buffer electrode (gain=1) using the same operational amplifier (LMC7111) as well as the spectral voltage noise of the total amplifier (including the second feedback loop). Different sampling rates were used above and below 100 Hz. f_0 denotes the upper bandwidth of the amplifying electrode defined by the gain-bandwidth product of the employed op-amp (LMC7111).

For comparison we added the spectral voltage-noise density of the same op-amp when used as active buffer electrode (gain of 1). The input-referred noise is higher even though the same op-amp is used. This confirms that using an amplifying electrode with a gain of about 40 dB lowers the noise when compared to the same active electrode configured as a buffer electrode. In addition, the noise of the subsequent stages will contribute less in the case of amplifying electrodes. The total noise of the buffer electrode in the bandwidth of 10 Hz to 1 kHz can be estimated to 11.1 μV_{rms} This improvement may be considered small, but it comes without cost. The noise reduction becomes more important the higher the gain of the active electrode is. To our knowledge 40 dB corresponds to the second highest gain[1] of any reported input stage of an amplifier for bioelectric events (excluding AC-coupled stages). This large gain has been possible because of the error-feedback loop which suppresses the electrode-electrolyte offset voltage.

Table 6.1 resumes the characteristics of the two amplifying electrodes compared to the buffer electrode presented in chapter 4.

---

[1] The highest reported gain of 43.5 dB was obtained using active electrodes with a stopped integrator in the feedback loop [Mett 97] but without any means to compensate for any gain mismatch.

6.3. Method

Tab. 6.1: The measured characteristics for both amplifying electrodes and the buffer electrode from chapter 4.

Electrodes LMC7111		amplifying electrode Fig. 6.1 a)	Fig. 6.1 b)	buffer electrode
I_{sup}	[mA]	1	1	1
Offset	[mV]	555 ± 25	3 ± 1	3 ± 1
VCC^a	[V]	≥ 4.45 V	≥ 3.9 V	≥ 3.7 V
$R_{in}{}^b$	[$G\Omega$]	5000	5000	5000
I_{bias}	[pA]	-60	-60	-60
R_{out}	[Ω]	65	< 0.1	< 0.1
Noise^c	[μV_{rms}]	9.3	9.3	11.1
Gain	[dB]	40.06	40.06	-0.01

^{a} VCC of the total amplifier assuming $V_{ref} = 500$ mV and the voltage drop over the current-source 1.2 V
^{b} Measured with a parameter analyzer HP4155B/4156B
^{c} total noise from 10 Hz to 1 kHz

For a high-quality system the noise has to be reduced by choosing the appropriate op-amps. We will show additional choices for op-amps in an addendum to these results later in section 6.4.

CMRR

To analyze the CMRR of a pair of electrodes of the amplifier in more detail we measured the two gains of these electrodes (quantifying the gain mismatch) and simulated the different gain functions in a P-Spice-based simulator (OrCAD). We also measured the CMRR of the total amplifier using these two electrodes in the frequency domain between 5 Hz and 50 kHz.

The different results of the total amplifier are shown in Fig. 6.9. The measured gains of the two electrodes allowed us to quantify the CMRR due to the gain mismatch as defined in equation (6.19) which led to a CMRR of about 30 dB. The simulation and measurements correspond to the case of a *non-compensated* gain mismatch and do not describe the performance of the final amplifier system.

Fig. 6.9 is an asymptotic representation of the different transfer functions defining the CMRR of one signal electrode 'i' in relation to the reference electrode 'r' of the amplifier. The simulation is based on two different spice simulations, one for the differential gain and one for the common-mode gain. The definitions of the different lines representing

110   6. Two-wired Amplifying Electrodes

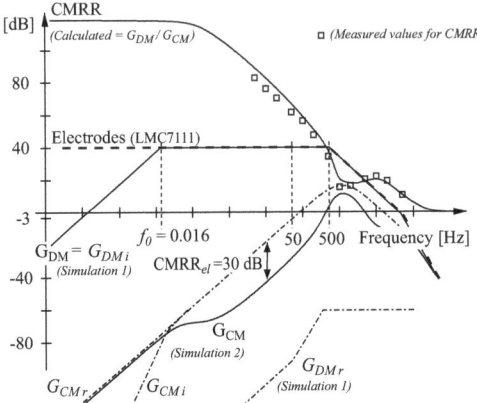

Fig. 6.9: The CMRR of the amplifier from Fig. 6.2 is defined by the ratio of the differential gain G_{DM} of the differential input signal divided by the common-mode gain G_{CM} of the differential input signal (here the two solid lines). The common-mode gain of the differential signal is affected by the difference in gain of the corresponding single-ended output signals. $G_{DM\,r}$ and $G_{CM\,r}$ are the differential and common-mode gain of the output voltage of the reference electrode $V_{out\,r}$, whereas $G_{DM\,i}$ and $G_{CM\,i}$ are the differential and common-mode gain of the output of the signal electrode 'i' (dotted lines). The dashed line 'Electrodes (LMC7111)' represents the gain of the active electrodes without the second feedback loop. The higher frequencies are suppressed as a result of the limited open-loop gain of the op-amp. The squares represent measured values of the CMRR for our prototype circuit. The value of $CMRR_{el}$ was limited by the mismatch in gain (@ 50 Hz) between the two electrodes as described in equation (6.19). The actual value of the CMRR between the two electrodes was only about 30 dB due to resistor tolerances.

the decomposition of the overall CMRR are compiled below:

$$G_{DM\,i} = \frac{V_{out\,i}}{V_{in\,i} - V_{in\,r}} \qquad G_{CM\,i} = 2\,\frac{V_{out\,i}}{V_{in\,i} + V_{in\,r}} \qquad (6.27)$$

$$G_{DM\,r} = \frac{V_{out\,r}}{V_{in\,i} - V_{in\,r}} \qquad G_{CM\,r} = 2\,\frac{V_{out\,r}}{V_{in\,i} + V_{in\,r}} \qquad (6.28)$$

$$G_{DM} = \frac{V_{out\,i} - V_{out\,r}}{V_{in\,i} - V_{in\,r}} \qquad G_{CM} = 2\,\frac{V_{out\,i} - V_{out\,r}}{V_{in\,i} + V_{in\,r}} \qquad (6.29)$$

As the gain differs for each signal electrode so does the CMRR of the amplifier for each pair of electrodes. This is a typical situation for a multi-electrode amplifier with non-coupled electrodes.

The differential-mode gain of the reference output voltage $G_{DM\,r}$ is very

6.3. Method

low because V_{out} is forced to a constant value by the first feedback loop. $G_{DM\,i}$ corresponds to the transfer function of the active electrodes (line LMC7111) but is damped at low frequencies ($< f_0$) by the second feedback loop. Because $G_{DM\,i} \gg G_{DM\,r}$, the differential-to-differential gain G_{DM} already drawn in Fig. 6.3 also covers the $G_{DM\,i}$ curve. The two common-mode gains $G_{CM\,r}$ and $G_{CM\,i}$ almost overlap. Their mismatch is a result of the resistor tolerances resulting in a reduced CMRR defined by equation (6.19). The CMRR was measured for this particular pair of electrodes at about 30 dB. This is a little bit better than the 28-dB worst-case CMRR expected from the tolerance of the resistors as estimated by equation (2.27). For frequencies below the corner frequency of the second feedback loop $f_0 \approx 0.016\,Hz$ we see that $G_{CM\,r}$ and $G_{CM\,i}$ diverge while they are very close for frequencies above the corner frequency. It follows that the differential common-mode gain G_{CM} is mostly 30 dB below the single common-mode gains but follows $G_{CM\,r}$ below f_0.

The corner frequency of the first feedback loop (DRL loop) is expressed by $1/2\pi R_{DRL} C_{DRL}$ yielding 53 Hz. This corresponds to the frequency where the gain of $G_{CM\,r}$ equals -3 dB.

To summarize we can express the CMRR at 50 Hz by:

$$\text{CMRR} \approx 20 \log \left(\frac{G_i + G_r}{2(G_i - G_r)} \cdot \left(1 + \frac{G_{DM}}{2\pi R_{DRL} C_{DRL} 50} \right) \right) \approx 30 + 40\,dB \quad (6.30)$$

For our prototype we measured a CMRR of 61.7 dB at 50 Hz (with the DRL) for a *non-compensated* gain mismatch. This is still good enough for a body-worn amplifier (see table 2.2) like the one we built, but not for a clinical application which typically uses mains-powered amplifiers.

Yet, the measured CMRR is below the estimated CMRR from equation (6.30) which is probably due to the gain of the amplifying electrodes being affected by the limited gain-bandwidth product of the employed op-amp. This affects both terms in equation (6.30) by reducing the overall gain of the electrodes as well as by increasing their individual gain mismatch. We think that a larger gain-bandwidth product would improve the CMRR of the amplifier.

The low CMRR is probably the main reason why amplifying electrodes are not encountered often in clinical EEG and ECG. We will now evaluate the gain adaptation described before in section 6.3.2. To do so we must first measure a real ECG (or EEG) with the existing amplifier.

ECG

We now record a three-lead ECG. The system was composed of one DRL electrode together with the active reference electrode and one signal electrode, which corresponds to the amplifier shown in Fig. 6.2. We estimate that an ECG is best suited to evaluate the quality of a bioelectric signal due to its distinctive shape. The ECG signal was recorded without any skin preparation by using disposable self-adhering pre-gelled electrodes as depicted in Fig. 6.6. The output voltages of the amplifying electrodes $V_{out\ r}$ and $V_{out\ i}$ were sampled with a differential preamplifier and an oscilloscope (ADA400A and TDS744 from Tektronix) without any gain adaptation. The corresponding ECG is shown in Fig. 6.10:

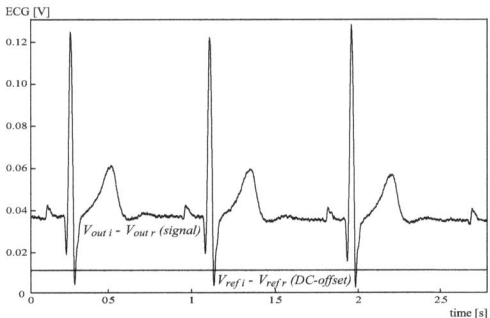

Fig. 6.10: A sample ECG recording made with the circuit shown in Fig. 6.2. The recording was made with disposable pre-gelled electrodes from Conmed. The gain of the active electrode was about 40 dB, the values of the resistors were $R_a = 47$ kΩ, $R_b = 9.1$ kΩ, $R_c = 4.7$ kΩ, $R_{fb} = 10$ MΩ, $C_{fb} = 10$ μF, $R_{DRL} = 220$ kΩ and $C_{DRL} = 15$ nF. The signal was sampled with an Oscilloscope making use of a differential preamplifier (TDS744 and ADA400A from Tektronix).

The ECG in Fig. 6.10 shows a flat baseline and very little 50-Hz interference. The DC voltage of the ECG should be maintained at zero by the second error-feedback loop. According to the figure, there is an offset of about 40 mV. This corresponds to the value calculated in equation (6.9).

The low-frequency parts of the signal were measured simultaneously ($V_{ref\ i} - V_{ref}$). The total DC offset between the electrodes is about 12 mV; supposedly the difference of the polarization voltage at the electrode-electrolyte interface (in contrast to the EEG there is no DC signal in an ECG).

To further improve the CMRR of the system we now implement the gain adaptation. For this reason we sample the individual signals using a differential ADC with an input range of 200 mV and 14-bit resolution (NI DAQCard-1200 from National Instruments) using the pseudo-differential configuration shown in Fig. 6.4 b). The resulting signals $V_{out\ i} - V_{DRL}$ and $V_{out} - V_{DRL}$ are displayed in Fig. 6.11:

6.3. Method

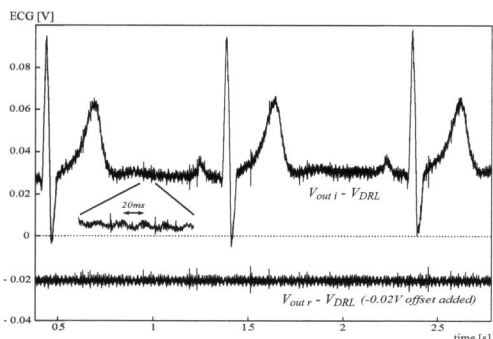

Fig. 6.11: A pseudo-differential ECG recording made with the pseudo-differential configuration shown in Fig. 6.4 b). The recording was made with a NI DAQCard-1200 from National Instruments. These two waveforms are normally not available with common amplifiers. The distinct noise visible on both waveforms corresponds to the common-mode signals which will be removed by calculating the difference similar to the signal shown in Fig. 6.10. For better visualization we added an artificial offset to $V_{out\,i} - V_{DRL}$ of -20 mV.

The two signals from Fig. 6.11 correspond to a pseudo-differential signal that is not normally accessible in standard amplifiers for bioelectric events. We can confirm the pseudo-differential nature of the configuration by the fact that the biopotential signal is present only on the signal electrode. However, the common mode interference is present on both signals (the visible 'noise'). Both signals still show all the common mode noise, e.g., the 50-Hz power-line interference is clearly visible in the enlargement. The also visible higher-frequency contributions are mostly induced by the oscilloscope (used for visualization) standing close to the subject. The frequency decomposition obtained by a Fourier transform of the two signals can be seen in the corresponding clipping; namely Fig. 6.12 a) and Fig. 6.12 b).

According to Fig. 6.12, the 50-Hz components of the two pseudo-differential signals are not the same. This is mainly due to the difference in the gain of the two electrodes, but also due to the fact that the bioelectric signal itself has a 50-Hz component which is only present on one lead; $V_{out\,i} - V_{DRL}$. Two factors are helpful in this situation. First, the 50-Hz common-mode interference is usually much larger than the 50-Hz component of the bioelectric signal. Second, the power-line interferences on the two signals are in phase. Thus, by taking the difference between the two unipolar signals the power-line interference will be vigorously reduced. Taking the difference results in a signal similar to the signal measured in Fig. 6.10 with the corresponding frequency decomposition

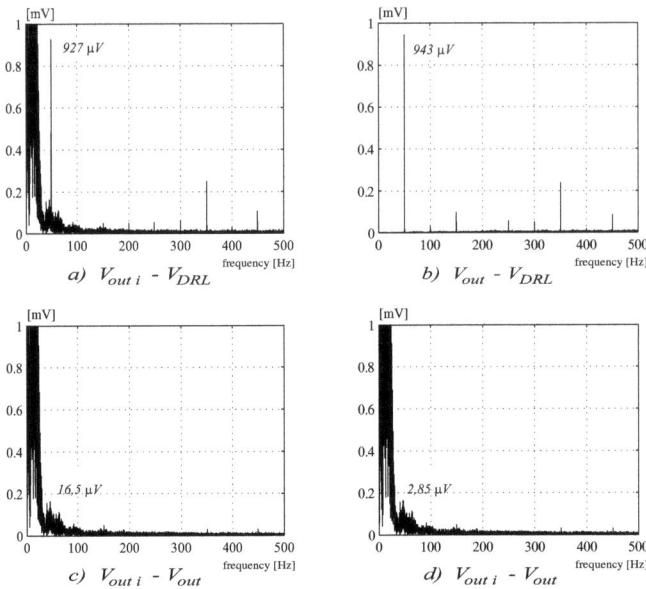

Fig. 6.12: The frequency decomposition of the two pseudo-differential signals as well as their difference with and without gain adaptation. Inset a) and b) are the two pseudo-differential signals shown in Fig. 6.11. Inset c) and d) are the digitally obtained differential signals. Inset c) corresponds to the simple difference of the two signals shown in Fig. 6.11, d) corresponds to the digitally improved signal (with gain adaptation) shown in Fig. 6.13. The value written in italic corresponds to the absolute value of the 50-Hz component. The FFTs correspond to a signal stretch of 4.5 s sampled with 1000 Hz.

shown in Fig. 6.12 c). From the figure we can estimate the CMRR from the gain mismatch to 35 dB. This value is close to the value of the first term given in equation (6.30). The existence of a 50-Hz component of the ECG signal which is not in phase with the power-line interference is confirmed by the fact that the difference of the two 50-Hz components (absolute value) of the two unipolar signals and the 50-Hz component of the difference signal do not match perfectly $(943 - 927 \leq 16.5)$.

We can go further and obtain a better reduction of the 50-Hz component by applying equation (6.20):

$$\text{ECG} = 2\,\frac{943\,V_{out\,i} - 927\,V_{out}}{943 + 927} \qquad (6.31)$$

Fig. 6.13 presents the resulting ECG and Fig. 6.12 d) shows the corre-

6.3. Method

sponding frequency decomposition.

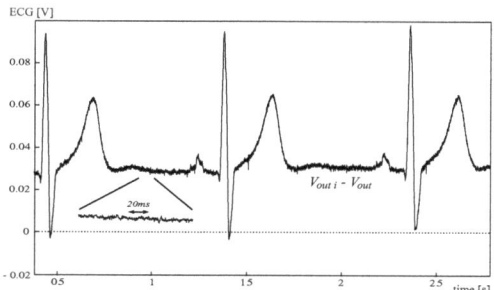

Fig. 6.13: The digitally calculated difference of the two pseudo-differential signals of Fig. 6.11. The enlargement shows no 50-Hz interference, the discrete Fourier transform in Fig. 6.12 d) reveals an amplitude of 2.85 μV of the 50-Hz component. This component is probably due to the bioelectric signal itself. The high-frequency spikes are probably due to the absence of any anti-aliasing filter in the experimental set-up.

The remaining 50-Hz component is about 2.85 μV. This is most probably not common mode interference but the bioelectric signal itself. This is supported by the fact that the phase between the 50-Hz component of the difference signal and the 50-Hz component of the reference signal is 88.7° which is nearly orthogonal. It is impossible to give an exact value for the CMRR, because we can only roughly estimate the common mode part and the difference mode part of the 50-Hz component. If the remaining 2.85 μV were only due to the differential signal, the CMRR at 50 Hz would be infinite. If we estimate that the part of the 2.85 μV being in phase with the 943 μV was the remaining common mode interference, then the improvement of the CMRR could be estimated to:

$$\Delta \mathrm{CMRR} = 20 \log \left(\frac{16.5}{2.85 \cos (88.7)} \right) \approx 48 \ dB \quad (6.32)$$

The total CMRR (including the 61.7 dB of the amplifier with DRL) would then amount to about 109.7 dB (@ 50 Hz).

It has to be noted that for an optimal CMRR the bandwidth of the amplifier and the amplifying electrode should extend to at least two decades above the 50 Hz. A large enough bandwidth will ensure that the mismatch in the upper low-pass corner frequency does not introduce a CMRR limitation similar to the mismatch in the highpass filter which was analyzed in section 2.2.4.

The trace of the digitally calculated ECG has no visible 50-Hz common mode interference, which can be confirmed by looking at the corresponding frequency spectrum in Fig. 6.12 d). There is visibly more noise than in the fully differential recording shown in Fig. 6.10. One reason is that by using two differential ADCs the noise of four input signals was adding up (due to a limitation of the experimental setup) whereas in the recording shown in Fig. 6.10 only two input signals contribute to the noise. The other reason may be that for this prototype circuit there was no anti-aliasing filter and therefore no reduction of high frequency noise. The different recordings were taken on different days and the electrodes were not placed at exactly the same location.

Digital Control Loop

We also implemented the digital control loop as depicted in Fig. 6.5. The resulting ECG is shown in Fig. 6.10.

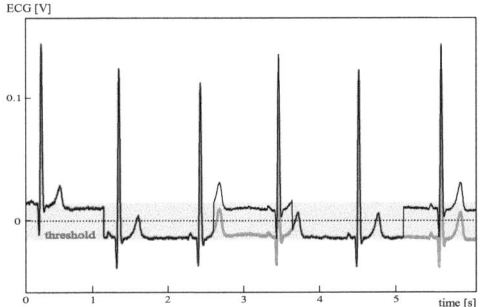

Fig. 6.14: A sample ECG recording obtained with a digital control loop as shown in Fig. 6.5. The baseline steps are a result of the discrete steps of the DAC. Their amplitude corresponds to the residual offset calculated in equation (6.25). The corner frequency of the digital highpass was calculated according to equation (6.24). The mean value of the recording is approximately 0 V. The oscillation can be suppressed by implementing a hysteresis.

As an example, the digital filter we used was implemented in LabView and consisted of a lowpass filter with a length of 511 tabs[1] followed by an integrator. The group delay (at 1000-Hz sampling frequency) was about 0.26 s. Because of the large group delay the additional delay through sampling and computing in LabView was not relevant.

The recording also demonstrates the presence of an oscillation (quantization noise) which is due to the discrete steps of the DAC. The observed step size of about 24 mV corresponds to the value calculated in equation (6.25). The gray line shows the recording when the software would in-

---
[1] this is the maximal number of tabs for a simple FIR filter in LabView

hibit any ADC change while the mean value of the bioelectric signal lies within a higher and a lower threshold.

The two thresholds are indicated by the semitransparent region behind the recordings. As we can see, there is also no superimposed hyperbolical relaxation curve like in highpass filtered signals. This enables a precise measurement of the S-T elevation even shortly after a spike which would have driven a conventional amplifier into saturation. Note, there was no capacitor required for the suppression of the DC offset.

6.4 Evaluation of Additional Op-Amps

We built the first amplifying electrodes using the LMC7111 (National Semiconductor). We later tested other op-amps for their suitability. The selection criteria were a high input impedance, low supply voltage (< 3 V) and low noise. All op-amps for low supply voltage which we found have a rail-to-rail input stage, although this is not required for the function of the amplifying electrode.

The start-up behavior cannot be simulated correctly. Therefore, we recommend always testing the op-amp by building an amplifying electrode and ramping up the input voltage from 0 V to the supply voltage of the amplifier two times. One time for a supply current which is below the intended supply current and one time for a supply current which is above the intended supply current. The op-amp has always to be tested with the final current-source.

Note, some op-amps do not require the diode $D1$ for proper functioning.

We found two additional op-amps which work seamlessly for the two-wired amplifying electrode, the LMV301 (National Semiconductors) and the TS1851 (Texas Instruments).

We then measured the spectral voltage-noise density of the active electrode with a gain of 40 dB ($R_1 = 330 \ \Omega$ and $R_2 = 33 \ \mathrm{k}\Omega$) using three different op-amps. The corresponding curves are shown in Fig. 6.15 together with the spectral voltage-noise density of an Ag/AgCl electrode [bios].

The total voltage noise of the electrode when using the LMV301 in the frequency band from 10 Hz to 1 kHz equals 5.4 μV_{rms}. The same figure for the electrode using the TS1851 equals 2.9 μV_{rms}. This is still larger than the total noise of an Ag/AgCl electrode which amounts to 1.5 μV_{rms}.

The low-frequency part is more critical and more difficult to estimate, but from fig 6.15 we can estimate that the spectral voltage-noise density of the active electrode built with the TS1851 is about the double of the

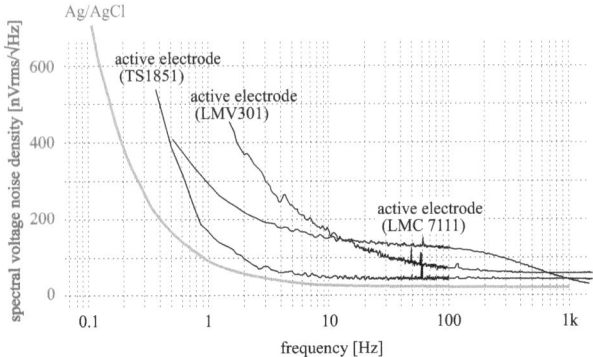

Fig. 6.15: The spectral voltage-noise density of the low-voltage two-wired amplifying electrodes was measured with a vector signal analyzer (Stanford SR785).

spectral voltage-noise density of the Ag/AgCl electrode. If we assume that the total voltage noise of the Ag/AgCl electrode integrated over the bandwidth from 0.1 Hz to 200 Hz equals 1.4 μV we than can estimate the total voltage noise of the active electrode over the same frequency band to about 2.8 μV.

This noise level is sufficiently low for the intended use in a portable ECG recorder because in that application the ECG is sampled with an 8-bit ADC only. Assuming a maximal peak of 1 mV for the ECG (see table 2.1) using an 8-bit ADC results in a resolution of about 4 μV. As a conclusion we state that the two-wired amplifying electrode based on the TS1851 is a good choice for portable ECG recorders.

Comparing again the two op-amps we see that they also differ in their input impedance. The LMV301 has a CMOS input stage and a rather high $1/f$ corner frequency. The TS1851 has lower noise level, but also a lower input impedance. The input stage of the TS1851 is not specified in the corresponding data sheet. The lower input impedance of the TS1851 will lead to a reduced CMRR due to the potential-divider effect (see section 2.2.4). Yet again, for portable applications the common mode requirement is lower (see table 2.2).

We also see that both additional op-amps have a larger gain-bandwidth-product (GBP) resulting in a larger bandwidth of the electrodes. This is very much welcome as the limited GBP of the LMC7111 may have limited the CMRR of the amplifier for bioelectric events (prior to gain adaptation)[1].

---

[1] As a general rule it is not recommended to relie on the limited bandwidth of an

Amplifying electrodes based on this three op-amps are compared in the Table 6.2.

Tab. 6.2: The characteristics according to the datasheet for all three op-amps when used for an amplifying electrode accordingly to Fig. 6.1 b).

Components		amplifying electrode		
		LMC7111	LMV301	TS1851
diodes required		1	1	-
I_{sup}	[mA]	1	1	1
VCC^a	[V]	$\geq$3.9 V	$\geq$3.5 V	$\geq$3.5 V
R_{in}	[GΩ]	5000	5000	N.A.
I_{bias}	[pA]	-60	-60	-10000
R_{out}	[Ω]	0.1	0.1	0.1
$\overline{v}_n{}^b$	[nV/$\sqrt{}$ Hz]	110	40	45
Noise^c	[μV$_{rms}$]	9.3	5.4	2.9
1/f corner	[Hz]	7	90	8
GBP	[kHz]	50	1000	480

^{a} assuming $V_{ref} = 500$ mV and the voltage drop over the current-source 1.2 V
^{b} spectral voltage-noise density at 1 kHz
^{c} measured total input-referred noise from 10 Hz to 1 kHz

6.5 Novelty

To the best of our knowledge, this is the first reported amplifier for bioelectric events which employs two-wired amplifying electrodes. The gain of the electrodes is about 40 dB. To prevent saturation of the output voltage of the electrodes a highpass filter is implemented. The filter is based on error feedback, thus preserving the high input impedance of the employed CMOS op-amps. To overcome the limitation of amplifying electrodes, namely the limitation of CMRR due to tolerance mismatch, the amplifier implements a gain adaptation which is done purely in software. To our knowledge this is the first time this gain adaptation is applied in software and not in hardware. This is only possible because of the pseudo-differential topology of the amplifier. Unlike a 50-Hz notch filter, the CMRR improvement achieved by our method reduces common mode interferences at all frequencies of interest.

Because the amplifier is able to compensate for a gain mismatch of the amplifying electrodes, and because the error-feedback loop can be implemented without the need of a large capacitor, this topology is also

op-amp to define the upper bandwidth of an amplifier.

suited for an integrated solution, e.g., this topology could be easily implemented in an ASIC. An example of an integrated amplifier for bioelectric signals taking advantage of an error feedback to dispose of the capacitors was published after this work by Rieger et al. [Rieg 09].

To our knowledge, this amplifier for bioelectric events was among the first amplifiers reported that are able to measure a physiological signal for frequencies down to DC. The measurement of the low-frequency components is achieved by sampling the error signal itself. This allows us to measure clinically relevant data normally not available with other reported amplifiers.

7. CONCLUSION

THIS thesis was motivated by the needs of the industry to develop small low-power amplifiers for bioelectric events intended for body-worn long-term-monitoring applications. The developed amplifiers are optimized for low-power consumption and a small form factor. They do not meet all the CE/FDA regulations for medical devices (e.g., concerning ESD protection and patient currents) as they are intended for monitoring applications only.

The results of this thesis influenced two devices which are used and sold today on the market, an electronic stethoscope and a belt-worn recorder for ECG, EEG and respiration. Beside the practical impact, four scientific journal publications emerged during this work.

In this chapter we will summarize the main results:

Electrodes

Throughout this work we tested various new materials for the design of electrodes. For hand-held devices we did not find a material which is more appropriate than the pre-gelled Ag/AgCl electrodes commonly used.

For long-term monitoring there are currently no electrodes available which satisfy all requirements. Pre-gelled electrodes lead to skin-irritation when applied for several hours. All other kinds of surface electrodes we tested or developed tend to suffer from bad ohmic contact, especially when used without a gel. We obtained very good results using conductive rubber electrodes which retain the moisture produced by the human body. The key for success might be keeping the electrodes wet during the recording, the EMPA (Swiss Federal Laboratories for Materials Science and Technology) in St. Gallen is currently developing such electrodes.

Active Electrodes

Active buffer electrodes are proven to reduce both motion artifacts and power-line interference. Their main disadvantages are the increased number of wires and the addition of noise to the total amplifier. The total noise could be kept low by using a simple FET-transistor as buffer electrode while employing low-noise non-FET op-amps in the subsequent stage.

Amplifying active electrodes have the potential to reduce the overall noise of an amplifier, but when being non-coupled, suffer from severe CMRR limitations. When coupled they require a large number of wires (typically five for signal electrodes).

For long-term monitoring we strongly recommend the use of dry active electrodes as it has been shown that they produce the best recordings. This is because pre-gelled electrodes tend to dry out when used over long periods of time (and they lead to skin irritation).

Active electrodes are an important part of this work. We contributed in two ways: First, we developed the first two-wired amplifying electrode for EEG measurements. Second, we developed two-wired buffer electrodes with very low noise which can be added to any existing amplifier and make it profit from the above mentioned benefits.

Amplifier Design

Amplifiers for biomedical applications require a low-noise input stage to maximize the signal-to-noise ratio (SNR) already limited by the inherent noise of the electrodes. In addition, they require a large common mode rejection ratio (CMRR) in order to successfully reduce the power-line interference. Part of the CMRR can be achieved by actively reducing the common-mode signal via a DRL circuit. However, the most important parameter of the amplifier remains the input impedance, which should be larger than $10^{12}\Omega$. A lower input impedance decreases the CMRR due to the potential-divider effect and increases motion artifacts.

Our contribution is manyfold: First, we developed a new circuit able to monitor the electrode-skin impedance mismatch present at the input of the amplifier without reducing the input impedance of the amplifier. This allows us to monitor the quality of the bioelectric recording (i.e., the contact of the electrodes) without degrading the amplifier in real time. Second, we demonstrated an autonomous method to compensate for any gain mismatch in non-coupled amplifying electrodes. Third, we developed a new pseudo-differential amplifier structure which allows us to compensate for a gain mismatch in software. The last two contribution both enable the application of non-coupled amplifying electrodes or digitizing electrodes without the severe reduction in CMRR this would mean in standard amplifiers. The error feedback employed to reject the electrode-electrolyte voltage used in the pseudo-differential amplifier can be implemented without the use of large capacitors which makes the described topology suitable for an integrated circuit (IC). In addition, it allows to measure the bioelectric signal down to DC, which opens the door to new applications.

Using a DRL Circuit

The use of a DRL circuit for amplifiers for bioelectric events is common practice. Although the design of DRL circuits is far from being a simple task there are not many publications to be found in scientific journals. There is only one article covering the analysis of a DRL circuit used with AC-coupled amplifiers [Burk 00]. One article shows the application of a DRL even for off-body measurements [Lim 06].

The use of a DRL circuit for amplifiers for bioelectric events is not challenged by the scientific community. Our own measurements show that the use of a DRL circuit may not be appropriate in the context of dry passive electrodes where a bad contact of either the ground electrode or the reference electrode leads to extensive clipping of the bioelectric signals. In this particular situation, the use of a DRL circuit leads to a loss of the biological information. We also showed that body-worn amplifiers may not require a DRL circuit due to the low powerline interference.

In all other circumstances we encourage the use of a DRL circuit.

Besides the reduction of the common-mode voltage, the DRL circuit can be used to force the common-mode voltage to a known voltage signal. The systems presented in this thesis are, to our knowledge, the first systems published which make use of the phase shift of this additional common-mode signal to control the common-mode voltage itself.

Using a 24-bit ADC

The use of 24-bit ADCs will increase in the future. For the time being, their power consumption is too large for many mobile applications or simply to large for an integrated solution. Their main advantage is that it becomes possible to use low-gain amplifier stages without any high-pass filter. This removes all clipping problems and thus preserves all information within the bioelectric recording.

Motion Artifacts

Motion artifacts can come from multiple origins. One is the capacitive coupling between wires. Another is the result of the potential-divider effect. Both of them can be reduced by proper design of the amplifier and also by the use of active electrodes.

In this work we claim that there is another source of motion artifacts which is dominant for hand-held devices. It is the intra-skin charge which is caused by very small currents inside the skin. It is this kind of motion artifact which is increased by a low input impedance of amplifiers for bioelectric events. We do not know of any non-invasive way to reduce this artifact. It is therefore important to minimize the weight and to mechanically decouple the electrodes from the remaining unit for wearable devices in order to minimize their variation of the contact pressure.

Outlook

The reduction of size and power consumption will continue and lead to more and more ubiquitous devices. Long-term monitoring of bioelectric signals will grow in importance as a proven instrument to ensure an optimal outcome of therapies, to increase the quality of life and decrease costs.

Only after the number of devices has grown as the number of electronic consumer devices has in the past, integrated solutions will become affordable. This will lead to even smaller devices which may be placed invisibly into our daily garments.

Future Research

During this thesis we developed a system consisting of two synchronized amplifiers for bioelectric events. Unfortunately this system was lost during the move of our laboratory. It would have been a perfect tool to undertake a comparative analysis using two different amplifiers independently recording the same bioelectric signal.

We would like to encourage researchers reading this thesis to rebuild such a system and to investigate the requirements of body-worn amplifiers in more detail. Such a system would, for example, allow one to quantify body-worn amplifiers in respect to the required CMRR. It would also allow quantifying the effect of a DRL circuit, or the use of polarizable electrodes on parameters like noise and motion artifacts. For this we would recommend to build electrodes with two interwoven but electrically isolated contact surfaces which then serve as input to two independent but synchronized amplifiers for bioelectric events.

Synopsis

Ein Herz schlägt mit etwa einem Hertz

ми # APPENDIX

A. FIGURES OF MERIT

Tab. A.1: Some important design parameters of the circuits described in the previous chapters.

Chapter / Circuit	R_{in} [Ω]	f_0 [Hz]	Gain [dB] first stage	$\bar{v}_n{}^a$ [$nV/\sqrt{Hz}$] @1 Hz	CMR [V]	CMRR^b [dB] @50 Hz
Chapter 2 Design Goals	^{c}> 10^{12}	0.016	> 20	< 70	±0.3	> 80
	^{d}> 10^{12}	0.016	> 14	^{e}<700	±0.15	> 20
Chapter 3 Impedance Mismatch	^{f}> 10^{13}	0.016	20	–	±0.25	–
Chapter 4 Buffer Electrode	^{g}> 10^{12}	N/A	−0.01	300	2.8 @5V	N/A
	^{h}> 10^{12}	N/A	−0.01	600	1.3 @5V	N/A
Chapter 5 Gain Adaption^j	^{i}> 10^{12}	0.016	29.8	–	±0.15	+23
	^{k}> 10^{12}	0.016	30.1	–	$v \pm 0.15$	+30
Chapter 6 Amplifying Electrode	^{l}> 10^{12}	0.016	40.06	300	±0.45 @5V	$^m \approx 30$
	^{n}> 10^{8}	0.016	–	170	–	–

[a] Input-referred spectral voltage-noise density
[b] when used with a DRL circuit
[c] high quality ECG amplifier
[d] wearable ECG amplifier
[e] This value was chosen by the author based on actual circuit performance. Value may change for different realizations.
[f] No change to underlying design
[g] NDS 336 @1mA
[h] LMC7111
[i] common-mode signal of 400 Hz
[j] No change to underlying design
[k] common-mode signal of 50 Hz
[l] LMC 7111
[m] CMRR increased to 60.7 dB with DRL and to 109.7 with gain adaptation
[n] TS 1851

B. SCHEMATICS FOR SIMULATION

The following sections show the schematics used for simulation and PCB design in OrCAD and served as a starting point for a development.

B.1 Two-Wired Amplifying Electrode

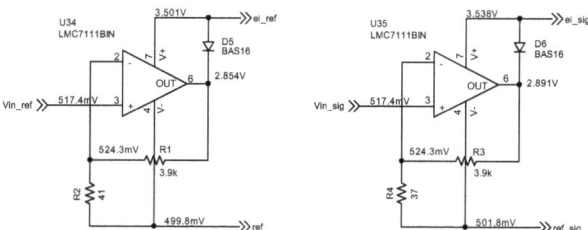

Fig. B.1: The schematic of the two-wired electrode used for simulation and PCB design.

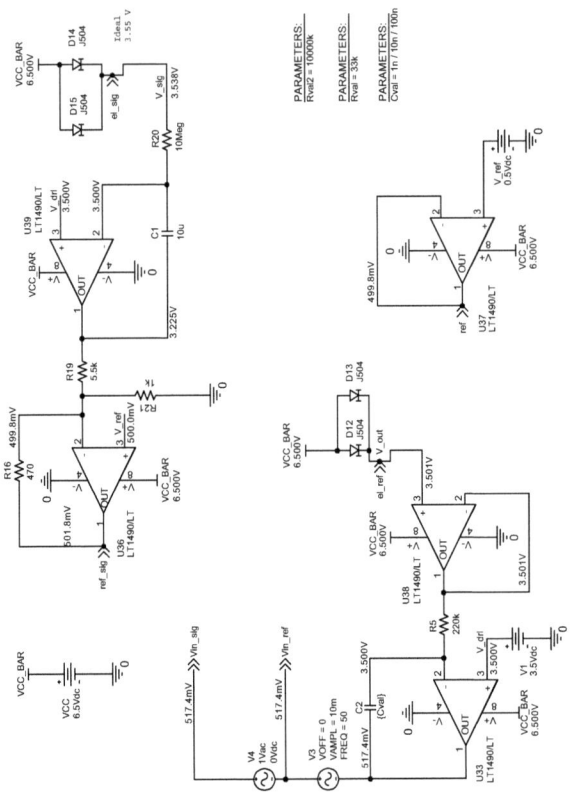

Fig. B.2: Second sheet of schematic used for simulation. The capacitor $C2$ in the DRL feedback loop is simulated for three different values, i.e., 1, 10 and 100 nF. There is one reference electrode and one signal electrode for simulation only.

B.1. Two-Wired Amplifying Electrode 133

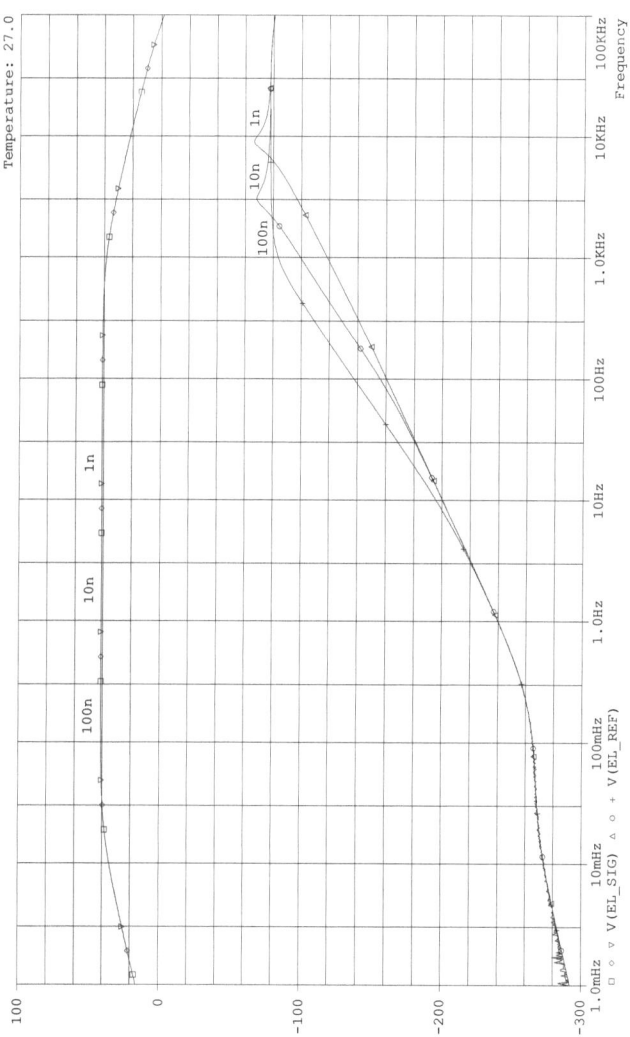

Fig. B.3: The simulated differential-mode gain for the two active electrodes, one used as reference electrode (in the DRL-loop) and one used as signal electrode for three distinct values of $C2$.

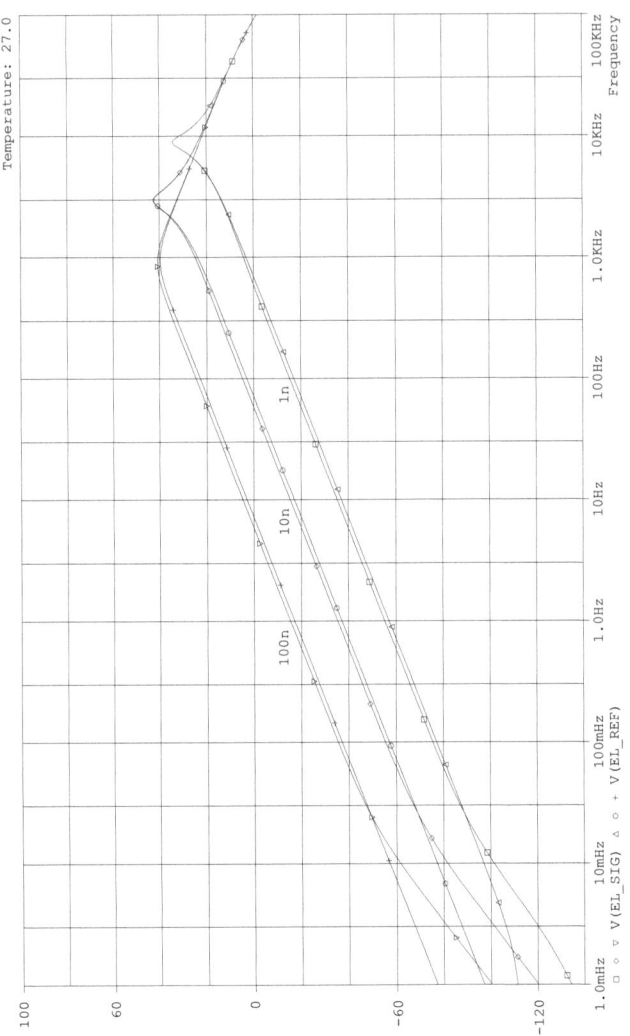

Fig. B.4: The simulated common-mode gain of the two active electrodes, one used as reference electrode (in the DRL-loop) and one used as signal electrode for three distinct values of $C2$. The two simulations are required to calculate the CMRR of the amplifier as depicted in Fig. 6.9.

B.2 Low-Voltage Two-Wired Amplifying Electrode

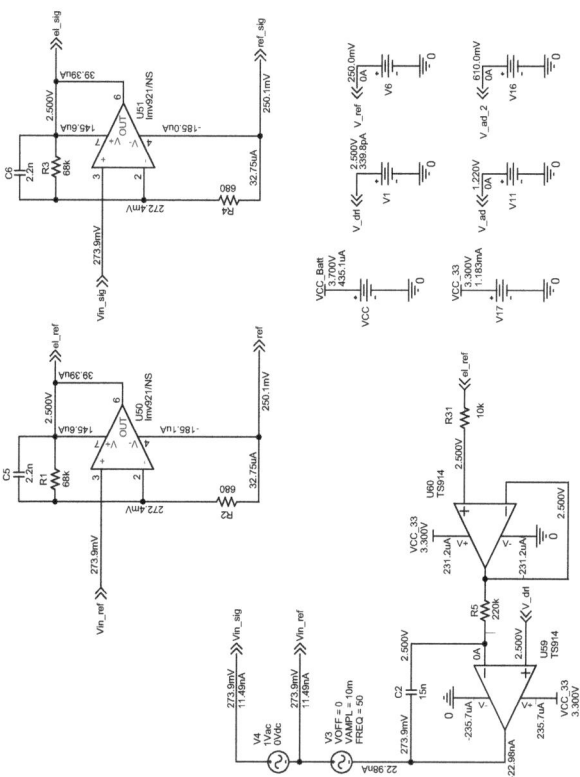

Fig. B.5: First sheet of schematic used for simulation. The two-wired electrodes have an additional capacitor which forms another lowpass filter for an increased overall filter order.

B. Schematics for Simulation

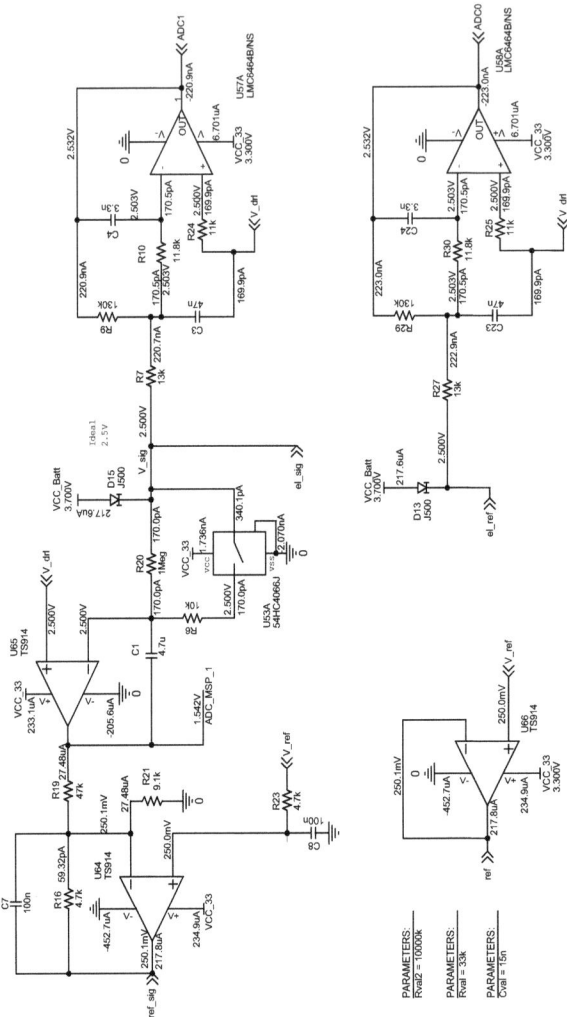

Fig. B.6: Second sheet of schematic used for simulation. There is one reference electrode and one signal electrode for simulation only. The two-wired electrode is followed by a second order multiple-feedback lowpass filter.

B.2. Low-Voltage Two-Wired Amplifying Electrode

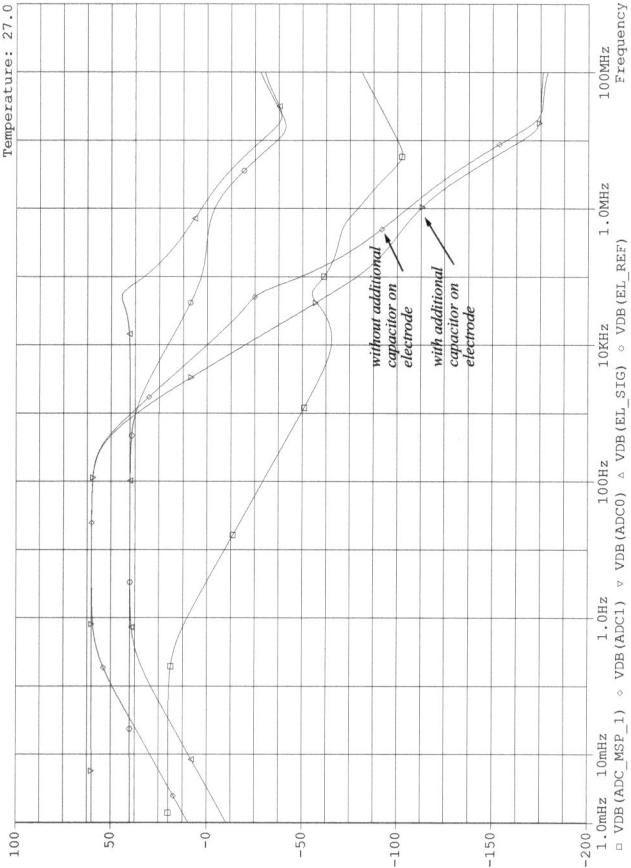

Fig. B.7: The simulated closed-loop gain of the two active electrodes after the electrodes and after the lowpass filter. The transfer function of the electrode is once shown with the additional capacitor and once without the additional capacitor for comparison. The low-frequency parts of the bioelectric signal is sampled as well ('ADC_MSP_1') but with a different ADC (at a lower sampling frequency).

B.3 Two-Wired Buffer Electrodes

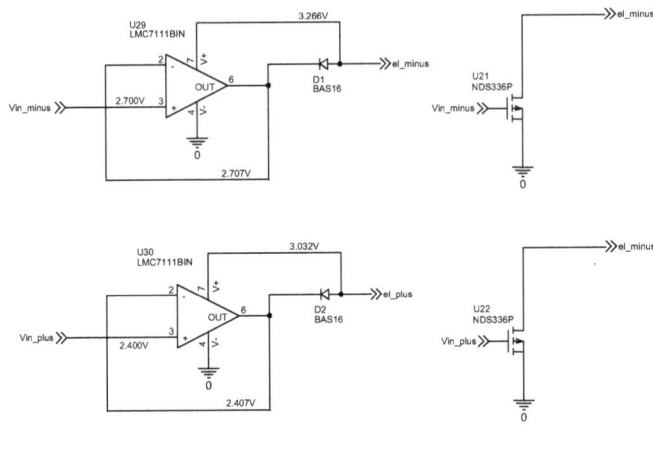

op-amp electrode   p-MOS FET electrode

Fig. B.8: The two-wired buffer electrodes exist in two versions. One employs an op-amp for buffering, one a simple p-MOS FET transistor. For simulation only one of the two pairs is connected (not shown in this schematic).

B.3. Two-Wired Buffer Electrodes

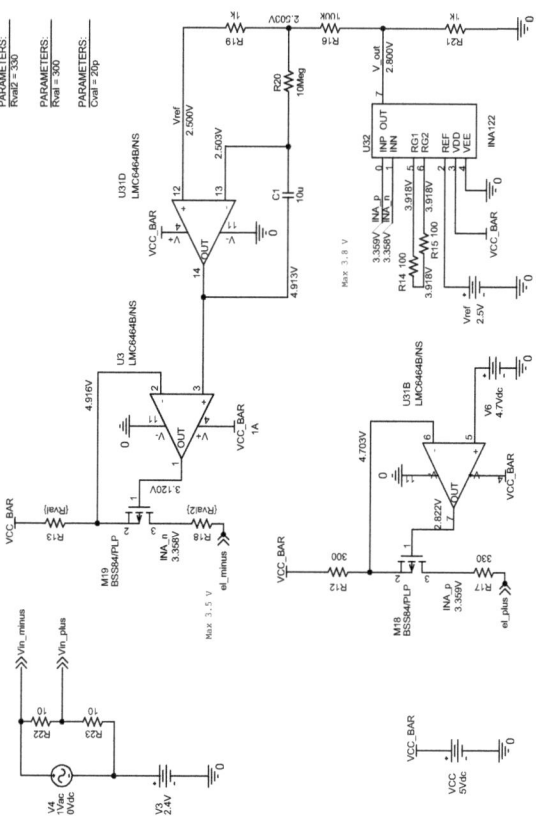

Fig. B.9: Second sheet of schematic used for simulation. There is one reference electrode and one signal electrode for simulation only. The current-source in this simulation is made with a op-amp and a transistor, this was used for investigating the influence of variances in the supply current. The AC source feeds a resistive divider in order to generate both a differential as a common-mode signal. This allows simulating the CMRR in one simulation.

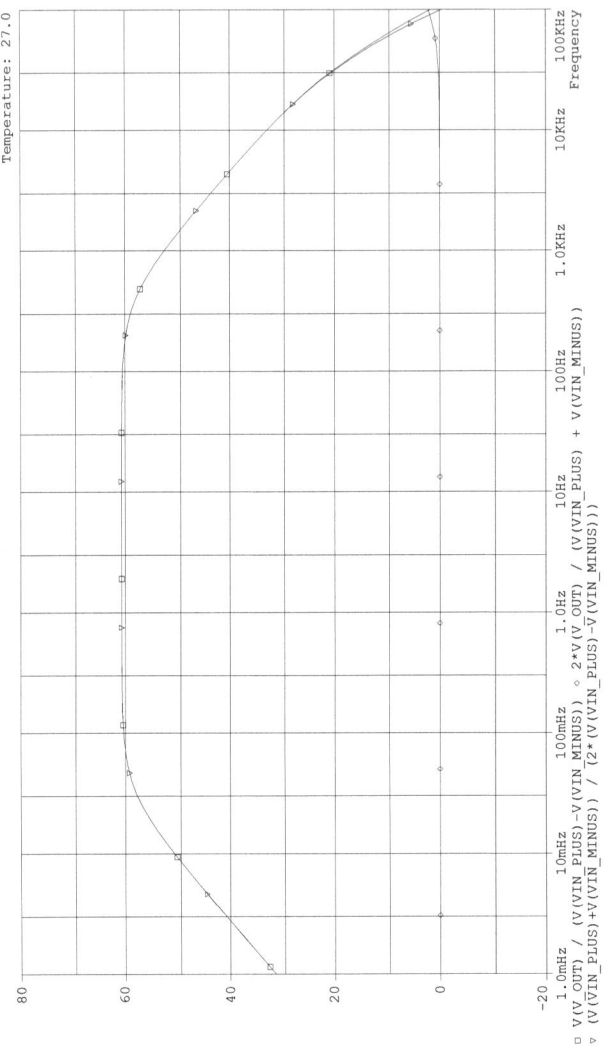

Fig. B.10: The simulated closed-loop common-mode gain and differential-mode gain of the amplifier with the two-wired buffer electrodes. The CMRR can be expressed by dividing the two expressions, this is only correct because the simulation tool uses the partial derivative.

BIBLIOGRAPHY

[e.g., 2004]   The bibliographic references are sorted in alphabetical order in three categories. In the first category are listed all publications related to biopotential measurements. The second category includes all references relating to methods and circuits for CMOS integrations. The last category encompasses all referred books, the latter references start with a four letter word put in *emphasis*.

Biopotentials

[Adli 98] X. Adli and Y. Yamamoto, "Impedance balancing analysis for power-line interference elimination in ECG signal," in *IMTC/98 Conference Proceedings*, 1998, vol. 1, pp. 235–238.

[Akin 99] T. Akin, B. Ziaie, S. A. Nikles, and K. Najafi, "A modular micromachined high-density connector system for biomedical applications," in *IEEE Transactions on Biomedical Engineering*, vol. 46, no. 4, pp. 471–480, 1999.

[Aliz 96] B. Alizadeh-Taheri, R. L. Smith, and R. T. Knight, "An active, microfabricated, scalp electrode array for EEG recording," in *Sensors and Actuators A (Physical)*, , no. 1, pp. 606–611, June 1996.

[Assa 09] C. Assambo, and M. J. Burke "Amplifier input impedance in dry electrode ECG recording," in *Proc. of the 31st Annual Intern. Conf. of the IEEE EMBS* Minneapolis, USA, pp. 1774–1777, Sep. 2009.

[Badi 00] L. Badillo, V. Ponomaryov, F. Gallegos, L. Igartua, and J. Gutierrez, "Long term multichannel EEG recorder digital system," in *Proceedings of the IASTED International Conference*, M. H. Hamza, Ed., 2000, pp. 275–278.

[Bell 00] A. Belardinelii, G. Palagi, R. Bedini, A. Ripoli, V. Macellari, and D. Franchi. "Advanced technology for personal biomedical signal logging and monitoring," in *Digest of Papers. Fourth International Symposium on Wearable Computers. IEEE Comput. Soc.*, 2000.

[Berd 02] L. Berdondini, T. Overstolz, F. de Rooij, N., M. Koudelka-Hep, S. Martinoia, P. Seitz, M. Wany, and N. Blanc, "High resolution electrophysiological activity imaging of in-vitro neuronal networks," in *2nd Annual International IEEE EMBS Special Topic Conference on Microtechnologies in Medicine and Biology*, A. Dittmar and D. Beebe, Eds., 2002.

[Berd 01] L. Berdondini, T. Overstolz, F. de Rooij, N., M. Koudelka-Hep, M. Wany, and P. Seitz, "High-density microelectrode arrays for electrophysiological activity imaging of neuronal networks," in *ICECS 2001*, A. Dittmar and D. Beebe, Eds., vol. 3, 2001.

[Berg 29] H. Berger, "Über das Elektrenkephalogramm des Menschen," in *European Arch. of Psychiatry and Clinical Neuroscience*, no. 87, pp. 527–570, 1929.

[Bin 95] H. Bin, Y. B. Chernyak, and R. J. Cohen, "An equivalent body surface charge model representing three-dimensional bioelectrical activity," in *IEEE Transactions on Biomedical Engineering*, vol. 42, no. 7, pp. 637–646, July 1995.

Bibliography 143

[Blum 03] R. A. Blum, J. D. Ross, C. M. Simon, E. A. Brown, R. R. Harrison, and S. P. DeWeerth, "A custom multielectrode array with integrated low-noise preamplifiers," in *Proceedings of the 25th Annual International Conference of the IEEE Engineering in Medicine and Biology Society IEEE Cat*, A. Dittmar and D. Beebe, Eds., 2003.

[Boni 95] S. Bonilla, L. Rasquinha, and P. Tarjan, "tripolar concentric ring electrodes for detecting forehead myoelectric potentials," in *1995 IEEE Engineering in Medicine and Biology 17th Annual Conference and 21 Canadian Medical and Biological Engineering Conference (Cat*,95, vol. 2, pp. 1549–1550, 1995.

[Brin 95] M. E. Brinson, and D. J. Faulkner, "New approaches ot measurement of operational amplifier common-mode rejection ration in the frequency domain," in *IEE Proc. Circuits Devices Systems*, vol. 142, no. 4, pp. 247–253, Aug. 1995.

[Burb 78] P. D. Burbank, and G. J. Webster, "Reducing skin potential motion artefact by skin abrasion," in *Medical & Biological Engineering & Computing*, vol. 16, pp. 31–38, Jan. 1978.

[Burk 00] M. J. Burke and D. T. Geeson, "A micropower dry-electrode ECG preamplifier," in *IEEE Transactions on Biomedical Engineering*, vol. 47, no. 2, pp. 155–162, Feb. 2000.

[Casa 94] O. Casas and R. Pallas-Areny, "Signal to noise ratio in bioelectrical impedance measurements using synchronous sampling," in *Proceedings of the 16th Annual International Conference of the IEEE Engineering in Medicine and Biology Society*, J. Sheppard-NF, M. Eden, and G. Kantor, Eds., vol. 2, pp. 890–891, 1994.

[Chim 00] M. F. Chimene and R. Pallas-Areny, "A comprehensive model for power line interference in biopotential measurements," in *IEEE Transactions on Instrumentation and Measurement*, vol. 49, no. 3, pp. 535–540, June 2000.

[Clel 81] A. D. McClellan, "Extracellular amplifier with bootstrapped input stage results in high common-mode rejection," in *Medical & Biological Engineering & Computing*, vol. 19, no. 5, pp. 657–658, Sept. 1981.

[Cros 84] W. G. Crosier and R. C. Lee, "Multipurpose EOG/ECG/EMG electrophysiological amplifier," in *Frontiers of Engineering and Computing in Health Care 1984*, J. L. Semmlow and W. Welkowitz, Eds., pp. 327–330, 1984.

[Crow 95] J. A. Crowe, A. Harrison, and B. R. Hayes-Gill, "The feasibility of long-term fetal heart rate monitoring in the home environment using maternal abdominal electrodes," in *Physiological Measurement*, vol. 16, no. 3, pp. 195–202, Aug. 1995.

[Dask 98]　　　I. K. Daskalov, I. A. Dotsinsky, and I. I. Christov, "Developments in ECG Acquisition, Preprocessing, Parameter Measurement, and Recording," in *IEEE Engineering in Medicine and Biology*, vol. 17, no. 2, pp. 50–58, March–April, 1998.

[Dege 03]　　　T. Degen, H. Jäckel, M. Rufer, and S. Wyss, "SPEEDY: a fall detector in a wrist watch," in *Proc. Seventh IEEE Intern. Symposium on Wearable Computers*, White Plains, USA, pp. 184–187, Oct. 2003.

[Dege 04]　　　T. Degen, and H. Jäckel, "Enhancing interference rejection of amplifying electrodes by automated gain adaption," in *IEEE Transactions on Biomedical Engineering*, vol. 51, no. 11, pp. 2031–2039, Nov. 2004.

[Dege 04b]　　T. Degen, and H. Jäckel, "Preamplified two-wired active electrodes with DC-offset compensation," in *Proc. of the 26th Annual Intern. Conf. of the IEEE EMBS*, San. Francisco, USA, pp. 2251–2254, Sep. 2004.

[Dege 06]　　　T. Degen, and H. Jäckel, "A pseudodifferential amplifier for bioelectric events with DC-offset compensation using two-wired amplifying electrodes," in *IEEE Transactions on Biomedical Engineering*, vol. 53, no. 2, pp. 300–310, Feb. 2006.

[Dege 07]　　　T. Degen, S. Torrent, and H. Jäckel, "Low-noise two-wired buffer electrodes for bioelectric amplifiers," in *IEEE Transactions on Biomedical Engineering*, vol. 54, no. 7, pp. 1328–1334, July 2007.

[Dege 08]　　　T. Degen, and H. Jäckel, "Continuous monitoring of electrode-skin impedance mismatch during bioelectric recordings," in *IEEE Transactions on Biomedical Engineering*, vol. 55, no. 6, pp. 1711–1714, June 2008.

[Devl 84]　　　P. H. Devlin, R. G. Mark, and J. W. Ketchum "Detection of electrode motion noise in ECG signals by monitoring electrode impedance," in *Proc. Computers in Cardiology, Los Angeles*, pp. 51–56, 1984.

[Dobr 05]　　　D. Dobrev, "Simple two-electrode biosignal amplifier," in *Medical & Biological Engineering & Computing*, vol. 43, pp. 725–729, 2005.

[Dobr 04]　　　D. Dobrev and I. Daskalov, "Two-electrode low supply voltage electrocardiogram signal amplifier," in *Medical & Biological Engineering & Computing*, vol. 42, pp. 272-276, 2004.

[Dobr 02]　　　D. Dobrev and I. Daskalov, "Two-electrode biopotential amplifier with current-driven inputs," in *Medical & Biological Engineering & Computing*, vol. 40, pp. 122-127, Jan. 2002.

[Dobr 02b] D. Dobrev, "Two-electrode non-differential biopotential amplifier," in *Medical & Biological Engineering & Computing*, vol. 40, pp. 546–549, Sep. 2002.

[Doku 99] Z. Dokur, T. Olmez, and E. Yazgan, "Comparison of discrete wavelet and fourier transforms for ECG beat classification," in *Electronics Letters*, vol. 35, no. 18, pp. 1502–1504, Sept. 1999.

[Dots 96] I. A. Dotsinsky, and I. K. Daskalov, "Accuracy of 50 hz interference subtraction from an electrocardiogram," in *Medical & Biological Engineering & Computing*, vol. 34, no. 6, pp. 489–494, Nov. 1996.

[Dots 91] I. A. Dotsinsky, I. I. Christov, and I. K. Daskalov, "Multichannel DC amplifier for a microprocessor electroencephalograph," in *Medical & Biological Engineering & Computing*, vol. 29, pp. 324–329, May 1991.

[Dots 85] I. A. Dotsinsky, I. I. Christov, C. L. Levkov, and I. K. Daskalov, "A microprocessor-electrocardiograph," in *Medical & Biological Engineering & Computing*, vol. 23, pp. 209–212, May 1985.

[Dozi 07] R. Dozio, A. Baba, C. Assambo, and M. J. Burke "Time based measurement of the impedance of the skin-electrode interface for dry electrode ECG recording," in *Proc. of the 29th Annual Intern. Conf. of the IEEE EMBS* Lyon, France, pp. 5001–5004, Sep. 2007.

[Duns 95] W. J. Ross Dunseath and E. F. Kelly, "Multichannel pc-based data-acquisition system for high-resolution EEG," in *IEEE Transactions on Biomedical Engineering*, vol. 42, no. 12, pp. 1212–1217, Dec. 1995.

[Eint 03] W. Einthoven, "Die galvanometrische Registrierung des menschlichen Elektrokardiogramms, zugleich eine Beurteilung der Anwendung des Capillarelektrometers in der Physiologie," in *Archiv für die gesammte Physiologie des Menschen und der Thiere*, no. 99, pp. 472–480, 1903.

[Enok 97] C. Enokawa, Y. Yonezawa, H. Maki, and M. Aritomo, "A microcontroller-based implantable telemetry system for sympathetic nerve activity and ECG measurement," in *Proceedings of the 19th Annual International Conference of the IEEE Engineering in Medicine and Biology Society*, H. K. Chang and Y. T. Zhang, Eds., vol. 5, pp. 2232–2234, 1997.

[Fern 00] M. Fernandez and R. Pallas-Areny, "Ag-AgCl electrode noise in high resolution ECG measurements," in *Biomedical Instrumentation & Technology*, vol. 34, no. 2, pp. 125–130, March 2000.

[Fern 99]   M. Fernandez and R. Pallas-Areny, "A comprehensive model for power-line interference in biopotential measurements," in *IMTC/99*, V. Piuri and M. Savino, Eds., vol. 1, pp. 573–578, 1999.

[Fern 97]   M. Fernandez and R. Pallas-Areny, "A simple active electrode for power line interference reduction in high resolution biopotential measurements," in *Proceedings of the 18th Annual International Conference of the IEEE Engineering in Medicine and Biology Society*, H. Boom, C. Robinson, W. Rutten, M. Neuman, and H. Wijkstra, Eds., vol. 1, pp. 97–98, 1997.

[Flic 00]   B. B. Flick and R. Orglmeister, "A portable microsystem-based telemetric pressure and temperature measurement unit," in *IEEE Transactions on Biomedical Engineering*, vol. 47, no. 1, pp. 12–16, Jan. 2000.

[Font 10]   M. B. A. Fontes, M. Op de Beeck, C. Van Hoof, and H. P. Neves "Tuning electrode impedance for the electrical recording of biopotentials," in *Proc. of the 32nd Annual Intern. Conf. of the IEEE EMBS* Buenos Aires, Argentina, pp. 1812–1815, Sep. 2010.

[Gagn 00]   S. Gagne, U. S. Ganguly, and S. Comtois, "Compensation of the differential floating capacitance between dual microelectrodes," in *IEEE Transactions on Biomedical Engineering*, vol. 47, no. 4, pp. 551–555, 2000.

[Gasu 00]   M. Gasulla, O. Casas, and R. Pallas-Areny, "On the common mode response of fully differential circuits," in *Proceedings of the 17th IEEE Instrumentation and Measurement Technology Conference [Cat*, V. Piuri and M. Savino, Eds., vol. 2, pp. 1045–1049, 2000.

[Godi 91]   D. T. Godin, P. A. Parker, and R. N. Scott, "Noise characteristics of stainless-steel surface electrodes," in *Medical & Biological Engineering & Computing*, vol. 29, no. 6, pp. 585–590, Nov. 1991.

[Gond 96]   C. Gondran, E Siebert, S. Yocoub, and E. Novokov, "Noise of surface bio-potential electrodes based on nasicon ceramic and Ag-AgCl," in *Medical & Biological Engineering & Computing*, vol. 34, no. 6, pp. 460–466, Nov. 1996.

[Gond 95]   C. Gondran, E. Siebert, P. Fabry, E. Novakov, and P. Y. Gumery, "Non-polarisable dry electrode based on nasicon ceramic," in *Medical & Biological Engineering & Computing*, vol. 33, no. 3, pp. 452–457, May 1995.

[Gorg 86]   A. Gorgenyi, B. Telkes, B. Pataki, and Z. Papp, "A multichannel biomedical amplifier system," in *Proceedings of the Eighth Annual Conference of the IEEE/Engineering in Medicine and Biology Society (Cat*, G. V. Kondraske and C. J. Robinson, Eds., vol. 2, pp. 750–753, 1986.

[Grim 97] C. A. Grimbergen, A. C. Metting van Rijn, A. P. Kuiper, R. H. Honsbeek, K. Speijer, and A. Peper, "Dc rejection and deblocking in multichannel bioelectric recordings," in *1995 IEEE Engineering in Medicine and Biology 17th Annual Conference and 21 Canadian Medical and Biological Engineering Conference (Cat*, vol. 2, pp. 1665–1666, 1997.

[Grim 94] C. A. Grimbergen, A. C. Metting van Rijn, and A. Peper, "System configurations with a/d conversion for multichannel bioelectric recordings," in *Proceedings of the 16th Annual International Conference of the IEEE Engineering in Medicine and Biology Society*, vol. 2, pp. 996–997, 1994.

[Grim 92] C. A. Grimbergen, A. C. Metting van Rijn, and A. Peper, "A method for the measurement of the properties of individual electrode-skin interfaces and the implications of the electrode properties for the amplifier," in *Proceedings of the 14 th Annual International Conference of the IEEE Engineering in Medicine and Biology Society*, vol. 6, pp. 2382–2383, 1992.

[Gris 02] P. Griss, H. K. Tolvanen-Laakso, P. Merilainen, and G. Stemme, "Characterization of micromachined spiked biopotential electrodes," in *IEEE Transactions on Biomedical Engineering*, vol. 49, no. 6, pp. 597–604, June 2002.

[Guy 93] C. N. Guy, S. Walker, G. Alarcon, C. D. Binnie, P. Chesterman, P. Fenwick, and S. Smith, "MEG and EEG in epilepsy: is there a difference?," in *Physiological Measurement*, vol. 14, Nov. 1993.

[Hack 04] J. R. Hackworth, "Measurement of Common Mode Rejection Ratio," in *Electrical Power and Machinery Laboratory, Navy College*, Example Formal Lab Report, http://www.lions.odu.edu/jhackwor/365nc/, last accessed Nov. 2004.

[Hage 85] B. Hagemann, G. Luhede, and H. Luczak, "Improved active electrodes for recording bioelectric signals in work physiology," in *European Journal of Applied Physiology*, vol. 54, pp. 95–98, 1985.

[Haje 99] M. F. Hajer, A. C. Metting Van Rijn, and C. A. Grimbergen, "Design of an optical power supply for biopotential measurement systems," in *Proceedings of the First Joint BMES/EMBS Conference*, H. Boom, C. Robinson, W. Rutten, M. Neuman, and H. Wijkstra, Eds., vol. 2, p. 845, 1999.

[Hami 96] P. S. Hamilton, "A Comparison of Adaptive and Nonadaptive Filters for Reduction of Power Line Interference in the ECG," in *IEEE Transactions on Biomedical Engineering*, vol. 43, no. 1, pp. 105–109, January 1996.

[Hami 00]   P. S. Hamilton, M. G. Curley, R. M. Aimi, abd C. Sae-Hau, "Comparison of Methods for Adaptive Removal of Motion Artifact," in *Computers in Cardiology*, vol. 27, pp. 383–386, 2000.

[Hams 84]   G. H. Hamstra, A. Peper, and C. A. Grimbergen, "Low-power, low-noise instrumentation amplifier for physiological signals," in *Medical & Biological Engineering & Computing*, vol. 22, no. 3, pp. 272–274, May 1984.

[Harl 02]   C. J. Harland, T. D. Clark, and R. J. Prance, "Electric potential probes - new directions in the remote sensing of the human body," in *Measurement Science and Technology*, vol. 13, pp. 163–169, 2002.

[Harl 02b]   C. J. Harland, T. D. Clark, and R. J. Prance, "Remote detection of human electroencephalograms using ultrahigh input impedance electric potential sensors," in *Applied Physics Letters*, vol. 81, no. 17, pp. 3284–3286, Oct. 2002.

[Harl 03]   C. J. Harland, T. D. Clark, and R. J. Prance, "High resolution ambulatory electrocardiographic monitoring using wrist-mounted electric potential sensors," in *Measurement Science and Technology*, vol. 14, pp. 923–928, 2003

[Harr 03]   R. R. Harrison, and C. Charles, "A low-power low-noise CMOS amplifier for neural recording applications," in *IEEE Journal of Solid-State Circuits*, vol. 38, no. 6, June 2003.

[He 92]   B. He and R. J. Cohen, "Body surface laplacian ECG mapping," in *IEEE Transactions on Biomedical Engineering*, vol. 39, no. 11, pp. 1179–1191, Nov. 1992.

[Heun 84]   R. van Heuningen, H. G. Goovaerts, and F. R. de Vries, "A low noise isolated amplifier system for electrophysiological measurements: basic considerations and design," in *Medical & Biological Engineering & Computing*, vol. 22, no. 1, pp. 77–85, Jan. 1984.

[Holm 74]   G. Holmer-N, "Isolation amplifier energized by ultrasound," in *IEEE Transactions on Biomedical Engineering*, vol. 21, no. 4, pp. 329–333, July 1974.

[Hors 98]   M. J. van-der Horst, A. C. Metting van Rijn, A. Peper, and C. A. Grimbergen, "High frequency interference effects in amplifiers for biopotential recordings," in *Proceedings of the 20th Annual International Conference of the IEEE Engineering in Medicine and Biology Society*, H. K. Chang and Y. T. Zhang, Eds., vol. 6, pp. 3309–3312, 1998.

[Huan 98]   Q. Huang, and M. Oberle, "A 0.5-mw passive telemetry ic for biomedical applications," in *IEEE Journal of Solid State Circuits*, vol. 33, no. 7, pp. 937–946, July 1998.

[Huig 02]	E. Huigen, A. Peper, and C. A. Grimbergen, "Investigation into the origin of the noise of surface electrodes," in *Medical & Biological Engineering & Computing*, vol. 40, no. 3, pp. 332–338, May 2002.
[Hwan 08]	I.-D. Hwang, and J. G. Webster, "Direct interference canceling for two-electrode biopotential amplifier" in *IEEE Transactions on Biomedical Engineering*, vol. 55, no. 11, pp. 2620–2627, 2008.
[Iked 99]	A. Ikeda et al., "Focal ictal direct current shifts in human epilepsy as studied by subdural and scalp recording," in *Brain*, vol. 5, no. 122, pp. 827–838, 1999.
[Ishi 93]	M. Ishijima, "Monitoring of electrocardiograms in bed without utilizing body surface electrodes," in *IEEE Transactions on Biomedical Engineering*, vol. 40, no. 6, pp. 593–594, 1993.
[Jova 98]	E. Jovanoc, T. Martin, and D. Raskovic. "Issues in wearable computing for medical monitoring applications: A case study of a wearable ECG monitoring device," in *Proc. of the 20th Annual International Conference of the IEEE Engineering in Medicine and Biology Society*, 3:1295–1298, 1998.
[Jage 97]	P. J. A. deJager, R. Peper, A. C. Metting van Rijn, and C. A. Grimbergen, "Suppression of low frequency effects of high frequency interference in bioelectrical recordings," in *Proceedings of the 18th Annual International Conference of the IEEE Engineering in Medicine and Biology Society*, H. Boom, C. Robinson, W. Rutten, M. Neuman, and H. Wijkstra, Eds., vol. 1, pp. 26–27, 1997.
[Jin 87]	J. Jin and G. Schoffa, "Problems with isolation amplifiers in computer-based multichannel ECG systems," in *Biomedizinische Technik*, vol. 32, no. 5, pp. 107–112, May 1987.
[Jord 95]	C. Jordan, C. Weller, C. Thornton, and D. E. F. Newton, "Monitoring evoked potentials during surgery to assess the level of anaesthesia," in *Journal of Medical Engineering & Technology*, vol. 19, no. 2, pp. 77–79, March 1995.
[Kang 08]	T.-H. Kang, C. R. Merritt, E. Grant, B. Pourdeyhimi, and H. T. Nagle, "Nonwoven fabric active electrodes for biopotential measurement during normal daily activity," in *IEEE Transactions on Biomedical Engineering*, vol. 55, no. 1, pp. 188–195, 2008.
[Kim 04]	K. K. Kim, Y. K. Lim, and K. S. Park, "The electrically non-contacting ECG measurement on the toilet seat using the capacitively-coupled insulated electrodes," in *Proc. of the 26th Annual Intern. Conf. of the IEEE EMBS*, San. Francisco, USA, pp. 2375–2378, Sep. 2004.

[Ko 98] W. H. Ko, "Active electrodes for EEG and evoked potential," in *Proceedings of the 20th Annual International Conference of the IEEE Engineering in Medicine and Biology Society*, H. K. Chang and Y. T. Zhang, Eds., vol. 4, pp. 2221–2224, 1998.

[Ko 71] W. H. Ko, E. T. Yon, E. Greenstein, J. Hynecek, and D. Conrad, "A micropower telemetry system with active electrodes," in *Digest of technical papers of the 1971 international solid state circuits conference*, L. Winner, Ed.,71, pp. 102–103, 1971.

[Ko 70] W. H. Ko, M. R. Neuman, R. N. Wolfson, and E. T. Yon, "Insulated active electrodes," in *IEEE Transactions on Industrial Electronics and Control Instrumentation*, vol. 17, no. 2, pp. 195–198, 1970.

[Lant 97] G. Lantz, C. M. Michel, R. D. Pascual-Marqui, L. Spinelli, M. Seeck, S. Seri, T. Landis, and I. Rosen, "Extracranial localization of intracranial interictal epileptiform activity using loreta (low resolution electromagnetic tomography)," in *Electroencephalography and Clinical Neurophysiology*, vol. 102, no. 5, pp. 414–422, May 1997.

[Laud 98] M. K. Laudon, J. G. Webster, R. Frayne, and T. M. Grist, "Minimizing interference from magnetic resonance imagers during electrocardiography," in *IEEE Transactions on Biomedical Engineering*, vol. 45, no. 2, pp. 160–164, Feb. 1998.

[Laug 94] J. A. McLaughlin, E. T. McAdams, and J. Anderson, "Novel dry electrode ECG sensor system," in *Proceedings of the 16th Annual International Conference of the IEEE Engineering in Medicine and Biology Society*, J. Sheppard-NF, M. Eden, and G. Kantor, Eds., vol. 2, p. 804, 1994.

[Levk 82] L. Levkov, C., "Amplification of biosignals by body potential driving," in *Medical & Biological Engineering & Computing*, vol. 20, pp. 248–250, March 1982.

[Li 96] J. H. Li, C. Joppek, and U. Faust, "Fast eit data acquisition system with active electrodes and its application to cardiac imaging," in *Physiological Measurement*, vol. 17, Nov. 1996.

[Libe 83] B. Libet, et al., "Time of Conscious Intention to Act in Relation to Onset of Cerebral Activity (Readiness-Potential). The Unconscious Initiation of Freely Voluntary Act," in *Brain*, vol. 106 (Pt 3), pp. 623–642, 1983.

[Lim 04] Y. K. Lim, K. K. Kim, and K. S. Park, "The ECG measurement in the bathtub using the insulated electrodes," in *Proc. of the 26th Annual Intern. Conf. of the IEEE EMBS*, San. Francisco, USA, pp. 2383–2385, Sep. 2004.

[Lim 06] Y. G. Lim, K. K. Kim, and K. S. Park, "ECG measurement on a chair without conductive contact," in *IEEE Transactions* on *Biomedical Engineering*, vol. 53, no. 5, pp. 956–959, May. 2006.

[Linn 95] A. C. Linnenbank, A. C. MettingVanRijn, C. C. Grimbergen, and J. M. T. DeBakker "Choosing the resolution in AD conversion of biomedical signals," in *Proc. of the XXIInd International Congress on Electrocardiology, Nijmegen, The Netherlands*, vol. 16, 198-199, 1995.

[Lu 97] C. C. Lu, R. Plourde, W. Y. Fang, S. Uhlhorn, and P. P. Tarjan, "Laplacian electrocardiograms with active electrodes," in *Proceedings of the 1997 16th Southern Biomedical Engineering Conference (Cat*, J. D. Bumgardner and A. D. Puckett, Eds., pp. 121–124, 1997.

[Lu 97b] C. C. Lu, W. Y. Feng, and P. P. Tarjan, "Laplacian electrocardiograms with active electrodes for arrhythmia detection," in *Proceedings of the 19th Annual International Conference of the IEEE Engineering in Medicine and Biology Society*, J. D. Bumgardner and A. D. Puckett, Eds., vol. 1, pp. 363–366, 1997.

[Maas 82] A. J. J. Maas and M. Rodenberg, "Digital automatic circuit for over-range correction," in *Medical & Biological Engineering & Computing*, vol. 20, no. 2, pp. 245–247, March 1982.

[Maki 98] H. Maki, Y. Yonezawa, E. Harada, and I. Ninomiya, "An implantable telemetry system powered by a capacitor having high capacitance," in *Proceedings of the 20th Annual International Conference of the IEEE Engineering in Medicine and Biology Society*, H. K. Chang and Y. T. Zhang, Eds., vol. 4, pp. 1943–1946, 1998.

[Mars 84] I. Marshall and J. M. Neilson, "Mains interference in ECG recording," in *Journal of Medical Engineering & Technology*, vol. 8, no. 4, pp. 177–180, July 1984.

[Mett 97] A. C. Metting van Rijn, A. P. Kuiper, T. E. Dankers, and C. A. Grimbergen, "Low-cost active electrode improves the resolution in biopotential recordings," in *Proceedings of the 18th Annual International Conference of the IEEE Engineering in Medicine and Biology Society*, H. Boom, C. Robinson, W. Rutten, M. Neuman, and H. Wijkstra, Eds., vol. 1, pp. 101–102, 1997.

[Mett 94] A. C. Metting van Rijn, A. Peper, and C. A. Grimberger, "Amplifiers for bioelectric events: a design with a minimal number of parts," in *Medical & Biological Engineering & Computing*, vol. 32, no. 3, pp. 305–310, May 1994.

[Mett 94b] A. C. Metting van Rijn, A. Peper, and C. A. Grimbergen, "A wireless infrared link for a 16-channel EEG telemetry system," in *Proceedings of the 16th Annual International Confer-*

ence of the IEEE Engineering in Medicine and Biology Society, J. Sheppard-NF, M. Eden, and G. Kantor, Eds., vol. 2, pp. 906–907, 1994.

[Mett 93] A. C. Metting van Rijn, A. P. Kuiper, A. C. Linnenbank, and C. A. Grimbergen, "Patient isolation in multichannel bioelectric recordings by digital transmission through a single optical fiber," in *IEEE Transactions on Biomedical Engineering*, vol. 40, no. 3, pp. 302–308, March 1993.

[Mett 91] A. C. Metting van Rijn, A. Peper, and C. A. Grimbergen, "High-quality recording of bioelectric events. part 2. low-noise, low-power multichannel amplifier design," in *Medical & Biological Engineering & Computing*, vol. 29, no. 4, pp. 433–440, July 1991.

[Mett 91b] A. C. Metting van Rijn, A. Peper, and C. A. Grimbergen, "The isolation mode rejection ratio in bioelectric amplifiers," in *IEEE Transactions on Biomedical Engineering*, vol. 38, no. 11, pp. 1154–1157, Nov. 1991.

[Mett 90] A. C. Metting van Rijn, A. Peper, and C. A. Grimbergen, "High-quality recording of bioelectric events. part 1. interference reduction, theory and practice," in *Medical & Biological Engineering & Computing*, vol. 28, no. 5, pp. 389–397, Sept. 1990.

[Mitr 82] T. K. Mitra, B. R. Das, S. C. Bera, and P. C. Dhara, "A novel design criterion of a bio-instrumentation solid-state pre-amplifier," in *Journal of the Institution of Electronics and Telecommunication Engineers*, vol. 28, no. 6, pp. 282–286, June 1982.

[Mohs 04] P. Mohseni and K. Najafi, "A fully integrated neural recording amplifier with dc input stabilization," in *IEEE Transactions on Biomedical Engineering*, May 2004.

[Mohs 02] P. Mohseni and K. Najafi, "A low power fully integrated bandpass operational amplifier for biomedical neural recording applications," in *Conference Proceedings*, A. Dittmar and D. Beebe, Eds., vol. 3, 2002.

[Mohs 03] P. Mohseni and K. Najafi, "A wireless fm multi-channel microsystem for biomedical neural recording applications," in *2003 Southwest Symposium on Mixed Signal Design Cat*, A. Dittmar and D. Beebe, Eds., 2003.

[Mora 98] F. Mora, G. Villegas, A. Hernandez, W. Conronado, R. Justiniano, and G. Passariello, "Design of an instrumentation system for a neurocardiology laboratory," in *Proceedings of the 1998 Second IEEE International Caracas Conference on Devices, Circuits and Systems*, pp. 272–277, 1998.

[Mund 00]	C. W. Mundt and H. T. Nagle, "Applications of spice for modeling miniaturized biomedical sensor systems," in *IEEE Transactions on Biomedical Engineering*, vol. 47, no. 2, pp. 149–154, Feb. 2000.
[Nish 93]	S. Nishimura and Y. Tomita, "Development of an active electrode with an amplifier for surface electromyogram," in *Transactions of the Society of Instrument and Control Engineers*, vol. 29, no. 12, pp. 1474–1476, 1993.
[Nish 92]	S. Nishimura, Y. Tomita, and T. Horiuchi, "Clinical application of an active electrode using an operational amplifier," in *IEEE Transactions on Biomedical Engineering*, vol. 39, no. 10, pp. 1096–1099, Oct. 1992.
[Ober 82]	T. Oberg, "A circuit for contact monitoring in electrocardiography," in *IEEE Transactions on Biomedical Engineering*, vol. 29, no. 5, pp. 361–364, May 1982.
[Oedm 89]	S. Ödman, "Changes in skin potentials induced by skin compression," in *Medical & Biological Engineering & Computing*, Vol. 27, pp. 390–393, July 1989.
[Oedm 82]	S. Ödman, "On the spread of deformation potentials in the skin," in *Medical & Biological Engineering & Computing*, Vol. 20, pp. 451–456, July 1982.
[Oedm 81]	S. Ödman, "Potential and impedance variations following skin deformation," in *Medical & Biological Engineering & Computing*, Vol. 19, pp. 271–278, May 1981.
[Ohya 99]	M. Ohyama, Y. Tomita, S. Honda, H. Uchida, and N. Matsuo, "Active wireless electrodes for multichannel surface electromyography," in *Transactions of the Institute of Electrical Engineers of Japan, Part E*, vol. 119, no. 5, pp. 279–284, May 1999.
[Ohya 97]	M. Ohyama, Y. Tomita, S. Honda, H. Uchida, and N. Matsuo, "Active wireless electrodes for surface electromyography," in *Proceedings of the 18th Annual International Conference of the IEEE Engineering in Medicine and Biology Society*, H. Boom, C. Robinson, W. Rutten, M. Neuman, and H. Wijkstra, Eds., vol. 1, pp. 295–296, 1997.
[Olss 03]	R.-H. Olsson III, A.-N. Gulari, and K.-D. Wise, "A fully-integrated bandpass amplifier for extracellular neural recording," in *Conference Proceedings*, 2003.
[Olss 02]	R.-H. Olsson III, M. N. Gulari, and K. D. Wise, "Silicon neural recording arrays with on-chip electronics for in-vivo data acquisition," in *2nd Annual International IEEE EMBS Special Topic Conference on Microtechnologies in Medicine and Biology*, A. Dittmar and D. Beebe, Eds., 2002.

[Olth 93]   W. Olthuis, A. Volanschi, J. G. Bomer, and P. Bergveld, "A new probe for measuring electrolytic conductance," in *Sensors and Actuators B (Chemical)*, , no. 1, pp. 230–233, May 1993.

[Padm 90]   F. Z. Padmadinata, J. J. Veerhoek, G. J. A. van Dijk, and J. H. Huijsing, "Microelectronic skin electrode," in *Sensors and Actuators B (Chemical)*, , no. 1, pp. 491–494, Jan. 1990.

[Pall 93]   R. Pallas-Areny and J. G. Webster, "Bioelectric impedance measurements using synchronous sampling," in *IEEE Transactions on Biomedical Engineering*, vol. 40, no. 8, pp. 824–829, Aug. 1993.

[Pall 93b]   R. Pallas-Areny and J. G. Webster, "Ac instrumentation amplifier for bioimpedance measurements," in *IEEE Transactions on Biomedical Engineering*, vol. 40, no. 8, pp. 830–833, Aug. 1993.

[Pall 91]   R. Pallas-Areny and J. G. Webster, "Common mode rejection ratio in differential amplifiers," in *IEEE Transactions on Instrumentation and Measurement*, vol. 40, no. 4, pp. 669–676, Aug. 1991.

[Pall 91b]   R. Pallas-Areny and J. G. Webster, "Common mode rejection ratio for cascaded differential amplifier stages," in *IEEE Transactions on Instrumentation and Measurement*, vol. 40, no. 4, pp. 677–681, Aug. 1991.

[Pall 90]   R. Pallas-Areny and J. G. Webster, "Composite instrumentation amplifier for biopotentials," in *Annals of Biomedical Engineering*, vol. 18, no. 3, pp. 251–262, 1990.

[Pall 89]   R. Pallas-Areny, J. Colominas, and J. Rosell, "An improved buffer for bioelectric signals," in *IEEE Transactions on Biomedical Engineering*, vol. 36, no. 4, pp. 490–493, 1989.

[Pall 89b]   R. Pallas-Areny and J. Colominas, "Differential mode interferences in biopotential amplifiers," in *Images of the Twenty First Century*, Y. Kim and F. A. Spelman, Eds., vol. 5, pp. 1721–1722, 1989.

[Pall 88]   R. Pallas-Areny, "Interference-rejection characteristics of biopotential amplifiers: A comparative analysis," in *IEEE Transactions on Biomedical Engineering*, vol. 35, no. 11, pp. 953–959, Nov. 1988.

[Pall 87]   R. Pallas-Areny, "On the simulation of real 50-60-hz electrical fields (biological application)," in *IEEE Engineering in Medicine and Biology Magazine*, vol. 6, no. 1, pp. 58, March 1987.

[Pall 86]   R. Pallas-Areny, "On the reduction of interference due to common mode voltage in two-electrode biopotential amplifiers," in *IEEE Transactions on Biomedical Engineering*, vol. 33, no. 11, pp. 1043–1046, Nov. 1986.

[Para 03]   R. Paradiso, A. Gemignani, E-P. Scilingo, and D. De-Rossi, "Knitted bioclothes for cardiopulmonary monitoring," in *Proceedings of the 25th Annual International Conference of the IEEE Engineering in Medicine and Biology Society IEEE Cat*, vol. 4, pp. 3720–3723, 2003.

[Para 03b] R. Paradiso, "Wearable health care system for vital signs monitoring," in *Conference Proceedings of the 4th International IEEE EMBS Special Topic Conference on Information Technology Applications in Biomedicine*, pp. 283–286, 2003.

[Pepe 90]   A. Peper, R. Jonges, C. A. Grimbergen, T. G. Losekoot, and J. Strackee, "Method for the computation of an accurate zero reference for ECG signals," in *Medical & Biological Engineering & Computing*, vol. 28, no. 2, pp. 105–112, March 1990.

[Pepe 83]   A. Peper and C. A. Grimbergen, "EEG measurement during electrical stimulation," in *IEEE Transactions on Biomedical Engineering*, vol. 30, no. 4, pp. 231–234, 1983.

[Pran 00]   R. J. Prance, A. Debray, T. D. Clark, H. Prance, M. Nock, C. J. Harland, and A. J. Clippingdale, "Ultra-low-noise electrical-potential probe for human-body scanning," in *Measurement Science and Technology*, 2000.

[Ramo 99]   J. Ramos, R. Pallas-Areny, and M. Tresanchez, "Multichannel front-end for low level instrumentation signals," in *Measurement*, vol. 25, no. 1, pp. 41–46, Jan. 1999.

[Rein 87]   W. N. Reining, W. J. Tompkins, and J. G. Webster, "An ambulatory cardiac output monitor," in *Proceedings of the Ninth Annual Conference of the IEEE Engineering in Medicine and Biology Society (Cat*, V. Piuri and M. Savino, Eds., vol. 2, pp. 916–918, 1987.

[Rich 68]   P. C. Richardson and F. K. Coombs, "New construction techniques for insulated electrocardiographic electrodes," in *Proceedings of the annual conference on engineering in medicine and biology, Vol*, M. H. Hamza, Ed., 1968.

[Rieg 09]   R. Rieger, Y-Y. Pan, "A high-gain asquisition system with very large input range," in *IEEE Transactions on Circuits and Systems I: Regular Papers*, vol. 56, no. 9, pp. 1921–129, Sep. 2009.

[Rieg 03]   R. Rieger, J. Taylor, A. Demosthenous, N. Donaldson, and P. J. Langlois, "Design of a low-noise preamplifier for nerve cuff electrode recording," in *IEEE Journal of Solid-State Circuits*, vol. 38, no. 8, Aug. 2003.

[Rose 88] J. Rosell, J. Colominas, P. Riu, R. Pallas-Areny, and J. G. Webster, "Skin impedance from 1 Hz to 1 MHz," in *IEEE Transactions on Biomedical Engineering*, vol. 35, no. 8, pp. 649–651, Aug. 1988.

[Sabe 01] E. E. Sabelman, D. Schwandt, and D. L. Jaffe. "The wamas (wearable accelerometric motion analysis system: combining technology development and research in human mobility," in *Conf. Intellectual Property in the VA: Changes, Challenges & Collaborations, Arlington, VA*, 2001.

[Schu 02] Y. Schutz, S. Weinsier, P. Terrier, and D. Durrer. "A new accelerometric method to assess the daily walking practice," in *International Journal of Obesity*, 26:111–118, 2002.

[Sear 00] A. Searle, and L. Kirkup, "A direct comparision of wet, dry and insulating bioelectric recording electrodes" in *Physiol. Meas.*, vol. 21, pp. 271–283, Jan., 2000.

[Segu 04] J. J. Segura-Juárez, D. Cuesta-Frau, L. Samblas-Penam and M. Aboy, "A Microcontroller-Based Portable Electrocardiograph Recorder" in *Transactions on Biomedical Engineering*, vol. 51, no. 9, pp. 1686–1690, Sep., 2004.

[Shan 84] T. M. R. Shankar and J. G. Webster, "Design of an automatically balancing electrical impedance plethysmograph," in *Journal of Clinical Engineering*, vol. 9, no. 2, pp. 129–134, June 1984.

[Smit 87] H. W. Smit, K. Verton, and C. A. Grimbergen, "A low-cost multichannel preamplifier for physiological signals," in *IEEE Transactions on Biomedical Engineering*, vol. 34, no. 4, pp. 307–310, 1987.

[Spin 06] E. M. Spinelli, M. A. Mayoski, and R. Pallas-Areny, "A practical approach to electrode-skin impedance unbalance measurement," in *IEEE Transactions on Biomedical Engineering*, vol. 53, pp. 1451–1453, Jul. 2006.

[Spin 05] E. M. Spinelli, and M. A. Mayoski, "Two-electrode biopotential measurements: Power line interference analysis," in *IEEE Transactions on Biomedical Engineering*, vol. 52, pp. 1436–1442, Aug. 2005.

[Spin 04] E. M. Spinelli, N. Martinez, M. A. Mayoski, and R. Pallas-Areny, "A novel fully differential biopotential amplifier with DC suppression," in *IEEE Transactions on Biomedical Engineering*, vol. 51, pp. 1444–1448, Aug. 2004.

[Spin 03] E. M. Spinelli and R. Pallas-Areny, "Ac-coupled front-end for biopotential measurements," in *IEEE Transactions on Biomedical Engineering*, vol. 50, pp. 391–395, March 2003.

[Spin 01]	E. M. Spinelli, N. H. Martinez, and M. A. Mayosky, "A single supply biopotential amplifier," in *Medical Engineering & Physics*, , no. 23, pp. 235–238, Apr. 2001.
[Spin 00]	E. M. Spinelli and M. A. Mayosky, "Ac coupled three op-amp biopotential amplifier with active dc suppression," in *IEEE Transactions on Biomedical Engineering*, vol. 47, no. 12, pp. 1616–1619, Dec. 2000.
[Spin 99]	E. M. Spinelli, N. H. Martinez, and M. A. Mayosky, "A transconductance driven-right-leg circuit," in *IEEE Transactions on Biomedical Engineering*, vol. 46, no. 12, pp. 1466–1470, Dec. 1999.
[Stan 01]	K. Stangel, S. Kolnsberg, D. Hammerschmidt, B. J. Hosticka, H. K. Trieu, and W. Mokwa, "A programmable intraocular cmos pressure sensor system implant," in *IEEE Journal of Solid State Circuits*, vol. 36, no. 7, pp. 1094–1100, July 2001.
[Tahe 94]	B. A. Taheri, R. T. Knight, and R. L. Smith, "A dry electrode for EEG recording," in *Electroencephalography and Clinical Neurophysiology*, vol. 90, no. 5, pp. 376–383, May 1994.
[Talh 96]	H. de Talhouet and G. Webster, J., "The origin of skin-stretch-caused motion artifacts under electrodes," in *Physiological Measurement*, May 1996.
[Tayl 83]	D. Tayler and R. Vincent, "Signal distortion in the electrocardiogram due to inadequate phase response," in *IEEE Transactions on Biomedical Engineering*, vol. 30, no. 6, pp. 352–356, June 1983.
[Thak 80]	N. V. Thakor and J. G. Webster, "Ground-free ECG recording with two electrodes," in *IEEE Transactions on Biomedical Engineering*, vol. 27, no. 12, pp. 699–804, Dec. 1980.
[Toaz 98]	A. L. Toazza, F. Mendes-de Azevedo, and J. M. Neto, "Microcontrolled system for measuring skin/electrode impedance in bioelectrical recordings," in *Proceedings of the 1998 Second IEEE International Caracas Conference on Devices, Circuits and Systems*, pp. 278–281, 1998.
[Tsun 04]	D. Tsunami, J. McNames, A. Colbert, S. Person, and R. Hammerschlag "Variable frequency bioimpedance instrumentation," in *Proceedings of the 26th Annual International Conference of the IEEE Engineering in Medicine and Biology Society*, pp. 2386–2389, September 2004.
[Valc 04]	E. S. Valchinov, and N. E. Pallikarakis, "An active electrode for biopotential recording from small localized bio-sources" in *Biomedical Engineering Online*, 3:25, July 2004.

[Vanh 05]   S. Vanhatalo, J. Voipo, and K. Kaila, "Full-band EEG (FbEEG): an emerging standard in electroenchephalography," in *Clinical Neurophysiology*, vol. 116, no. 1 , pp. 1-8, Jan. 2005

[Vanh 02]   S. Vanhatalo et al., "DC-EEG discloses prominent, very slow activity patterns during sleep in preterm infants," in *Clinical Neurophysiology*, vol. 113, no. 11 , pp. 1822-1825, Nov. 2002

[Varl 96]   A. R. Varlan, and W. Sansen, "Characterisation of planar electrodes realised in planar microelectronic technology," in *Medical & Biological Engineering & Computing*, vol. 34, no. 4, pp. 308–312, July 1996.

[Verb 94]   M. Verbeke, S. Gur, A. Ron, and R. Iraqi, "Manual or automatic nulling DC offset for physiological DC amplifier," in *Medical Engineering & Physics*, vol. 16, pp. 171–171, March 1994.

[Webs 84]   J. G. Webster, "Reducing motion artifacts and interference in biopotential recording," in *IEEE Transactions on Biomedical Engineering*, vol. 31, no. 12, pp. 823–826, Dec. 1984.

[Webs 77]   J. G. Webster, "Interference and motion artifact in biopotentials," in *IEEE 1977 Region 6 Conference Record 'Electronics serving Mankind'*, V. Piuri and M. Savino, Eds., pp. 53–64, 1977.

[Whit 87]   D. R. White, "Phase compensation of the three op amp instrumentation amplifier," in *IEEE Transactions on Instrumentation and Measurement*, vol. 36, no. 3, pp. 842–844, Sept. 1987.

[Wint 83]   B. B. Winter and J. G. Webster, "Reduction of interference due to common mode voltage in biopotential amplifiers," in *IEEE Transactions on Biomedical Engineering*, vol. 30, no. 1, pp. 58–61, Jan. 1983.

[Wint 83b] B. B. Winter and J. G. Webster, "Driven-right-leg circuit design," in *IEEE Transactions on Biomedical Engineering*, vol. 30, no. 1, pp. 62–66, Jan. 1983.

[Woo 92]   E. J. Woo, P. Hua, J. G. Webster, W. J. Tompkins, and R. Pallas-Areny, "Skin impedance measurements using simple and compound electrodes," in *Medical & Biological Engineering & Computing*, vol. 30, no. 1, pp. 97–102, Jan. 1992.

[Yaco 97]   S. Yacoub, E. Novakov, Y. Gumery-P, C. Gondran, and E. Siebert, "Noise analysis of nasicon ceramic dry electrodes," in *1995 IEEE Engineering in Medicine and Biology 17th Annual Conference and 21 Canadian Medical and Biological Engineering Conference (Cat*, vol. 2, pp. 1553–1554, 1997.

[Yama 77]   T. Yamamoto and Y. Yamamoto, "Analysis for the change of skin impedance," in *Medical & Biological Engineering & Computing*, vol. 15, no. 3, pp. 219–227, May 1977.

[Zipp 79]   P. Zipp and E. Schad, "Quantification of motion artifacts in surface applied bioelectrodes," in *Biomedizinische Technik*, vol. 24, pp. 76–81, April 1979.

[Zipp 79b]   P. Zipp and H. Ahrens, "A model of bioelectrode motion artefact and reduction of artefact by amplifier input stage design," in *Journal of Biomedical Engineering*, vol. 1, no. 4, pp. 273–276, Oct. 1979.

[Zipp 78]   P. Zipp, H. Gessner, and H. Ahrens, "Methods of reducing movement interference when recording bioelectric signals by means of surface electrodes," in *Biomedizinische Technik*, vol. 23, pp. 1–2, May-June 1978.

CMOS integration

[Alza 01]   H. Alzaher and M. Ismail, "A cmos fully balanced differential difference amplifier and its applications," in *IEEE Transactions on Circuits and Systems II: Analog and Digital Signal Processing*, vol. 48, no. 6, pp. 614–620, June 2001.

[Bult 90]   K. Bult, and G. Geelen, "A fast settling CMOS op amp for SC circuits with 90 dB DC gain," in *IEEE J. Solid-State Circuits*, vol. 25, no. 6, pp. 1379–1384, Dec. 1990.

[Bund 00]   A. Bunde, S. Havlin, J. W. Kantelhardt, T. Penzel, H. Peter-J, and K. Voigt, "Correlated and uncorrelated regions in heart-rate fluctuations during sleep," in *Physical Review Letters*, vol. 85, no. 17, pp. 3736–3739, Oct. 2000.

[Chi 98]   Chi-Hung-Lin, and M. Ismail, "A low voltage CMOS rail-to-rail class-AB input/output opamp with slew-rate and settling enhancement," in *Proceedings of 1998 IEEE International Symposium on Circuits and Systems*, vol. 1, pp. 448–450, 1998.

[Chun 97]   H. Chung-Chih, M. Ismail, K. Halonen, and V. Porra, "Low-voltage rail-to-rail cmos differential difference amplifier," in *Proceedings of 1997 IEEE International Symposium on Circuits and Systems*, vol. 1, pp. 145–148, 1997.

[Degr 82]   M. G. Degrauwe, J. Rijmenants, E. A. Vittoz, and H. J. de Man, "Adaptive biasing CMOS Amplifiers," in *IEEE Journal of Solid-State Circuits*, vol. 17, no. 3, pp. 522–528, June 1982.

[Dezh 01]   Y. Dezhong, "A method to standardize a reference of scalp EEG recordings to a point at infinity," in *Physiological Measurement*, vol. 22, no. 4, pp. 693–711, Nov. 2001.

[Elwa 00]   H. Elwan, W. Gao, R. Sadkowski, and M. Ismail, "CMOS low-voltage class-AB operational transconductance amplifier," in *Electronics Letters*, vol. 36, no. 17, pp. 1439–1440, Aug. 2000.

[Hamm 98]   C. M. Hammerschmied and H. Qiuting, "Design and implementation of an untrimmed mosfet-only 10-bit a/d converter with -79-db thd," in *IEEE Journal of Solid State Circuits*, vol. 33, no. 8, pp. 1148–1157, Aug. 1998.

[Hanw 99]   L. Hanwoo and K. M. Buckley, "ECG data compression using cut and align beats approach and 2-d transforms," in *IEEE Transactions on Biomedical Engineering*, vol. 46, no. 5, pp. 556–564, May 1999.

[Huij 95]   J. H. Huijsing, R. Hogervorst, and K.-J. de Langen, "low-power low-voltage VLSI operational amplifier cells," in *IEEE Transanctions on Circuits and Systems I*, vol. 42, no. 11, pp. 841–852, Nov. 1995.

[Isma 99]   A. M. Ismail and A. M. Soliman, "Novel cmos linearised balanced output transconductance amplifier based on differential pairs," in *Frequenz*, vol. 53, no. 7, pp. 170–174, July 1999.

[Jane 92]   R. Jane, P. Laguna, N. V. Thakor, and P. Caminal, "Adaptive baseline wander removal in the ECG: Comparative analysis with cubic spline technique," in *Proceedings of Computer in Cardiology 1992 (Cat*, pp. 143–146, 1992.

[Kada 99]   S. Kadambe, R. Murray, and G. F. Boudreaux-Bartels, "Wavelet transform-based qrs complex detector," in *IEEE Transactions on Biomedical Engineering*, vol. 46, no. 7, pp. 838–848, July 1999.

[Kala 95]   T. Kalayci and O. Ozdamar, "Wavelet preprocessing for automated neural network detection of EEG spikes," in *IEEE Engineering in Medicine and Biology Magazine*, vol. 14, no. 2, pp. 160–166, March 1995.

[Kham 00]   A. Khamene and S. Negahdaripour, "A new method for the extraction of fetal ECG from the composite abdominal signal," in *IEEE Transactions on Biomedical Engineering*, vol. 47, no. 4, pp. 507–516, 2000.

[Klum 00]   E. A. M. Klumperink, S. L. J. Gierkink, A. P. van-der Wel, and B. Nauta, "Reducing mosfet 1/f noise and power consumption by switched biasing," in *IEEE Journal of Solid State Circuits*, vol. 35, no. 7, pp. 994–1001, July 2000.

[Kyun 00]   K. Kyung-Hwan and K. Sung-June, "Noise performance design of cmos preamplifier for the active semiconductor neural probe," in *IEEE Transactions on Biomedical Engineering*, vol. 47, no. 8, pp. 1097–1105, Aug. 2000.

[Lang 98]   K.-J. de Langen, and J. H. Huijsing, "Compact low-voltage power-efficient operational amplifier cells for VLSI," in *IEEE J. Solid-State Circuits*, vol. 33, no. 10, pp. 1482–1496, Oct. 1998.

[Loel 99]   T. Loeliger, and W. Guggenbühl, "Cascode configurations for switched current copiers," in *Analog Integrated Circuits and Signal Processing*, vol. 19, pp. 115–127, May 1999.

[Lu 98]   G. N. Lu and G. Sou, "1.3 v single-stage cmos opamp," in *Electronics Letters*, vol. 34, no. 22, pp. 2073–2074, Oct. 1998.

[Mart 98]   R. Martins, S. Selberherr, and F. A. Vaz, "A CMOS IC for Portable EEG Acquisition Systems" in *IEEE Transactions on Instrumentation and Measurement*, vol. 47, no. 5, pp. 1191–1196, Oct. 1998.

[Meno 99]   C. Menolfi and H. Qiuting, "A fully integrated, untrimmed cmos instrumentation amplifier with submicrovolt offset," in *IEEE Journal of Solid State Circuits*, vol. 34, no. 3, pp. 415–420, March 1999.

[Moro 01]   D. V. Morozov and A. S. Korotkov, "A realization of low-distortion cmos transconductance amplifier," in *IEEE Transactions on Circuits and Systems I: Fundamental Theory and Applications*, vol. 48, no. 9, pp. 1138–1141, Sept. 2001.

[Mort 00]   S. Mortezapour and E. K. F. Lee, "A 1-v, 8-bit successive approximation adc in standard cmos process," in *IEEE Journal of Solid State Circuits*, vol. 35, no. 4, pp. 642–646, 2000.

[Qing 00]   B. Qing, K. D. Wise, and D. J. Anderson, "A high-yield microassembly structure for three-dimensional microelectrode arrays," in *IEEE Transactions on Biomedical Engineering*, vol. 47, no. 3, pp. 281–289, March 2000.

[Qing 97]   B. Qing, K. D. Wise, J. F. Hetke, and D. J. Anderson, "A microassembly structure for intracortical three-dimensional electrode arrays," in *Proceedings of the 18th Annual International Conference of the IEEE Engineering in Medicine and Biology Society*, H. Boom, C. Robinson, W. Rutten, M. Neuman, and H. Wijkstra, Eds., vol. 1, pp. 264–265, 1997.

[Reay 95]   R. J. Reay, and T. A. Kovacs, "An unconditionally stable two-stage CMOS amplifier," in *IEEE Journal of Solid State Circuits*, vol. 30, no. 5, pp. 591–594, May 1995.

[Saec 90]   E. Säckinger, and W. Guggenbühl, "A highswing, high-impedance MOS cascode circuit" in *IEEE Journal of Solid-State Circuits*, vol. 25, pp. 289-Ü298, Feb.1990.

[Saha 00]   J. S. Sahambi, S. N. Tandon, and R. K. P. Bhatt, "An automated approach to beat-by-beat qt-interval analysis," in *IEEE Engineering in Medicine and Biology Magazine*, vol. 19, no. 3, pp. 97–101, May 2000.

[Senh 95]   L. Senhadji, G. Carrault, J. J. Bellanger, and G. Passariello, "Comparing wavelet transforms for recognizing cardiac patterns," in *IEEE Engineering in Medicine and Biology Magazine*, vol. 14, no. 2, pp. 167–173, March 1995.

[Shu 94] H. Shu-Chuan and M. Ismail, "Design of a cmos differential difference amplifier and its applications in a/d and d/a converters," in *APCCAS '94*, 1994, pp. 478–483, 1994.

[Tae 00] Y. Tae-Hwan, H. Fun-Jung, S. Dong-Yong, P. Se-Ik, O. Seung-Jae, J. Sung-Cherl, S. Hyung-Cheul, and K. Sung-June, "A micromachined silicon depth probe for multichannel neural recording," in *IEEE Transactions on Biomedical Engineering*, vol. 47, no. 8, pp. 1082–1087, Aug. 2000.

[Wang 90] Z. Wang, "Making cmos ota a linear transconductor," in *Electronics Letters*, vol. 26, no. 18, pp. 1448–1449, Aug. 1990.

[Wang 90b] Z. Wang and W. Guggenbuhl, "A voltage-controllable linear mos transconductor using bias offset technique," in *IEEE Journal of Solid State Circuits*, vol. 25, no. 1, pp. 315–317, Feb. 1990.

Books

[AAMI 99] American National Standard ANSI/AAMI EC38:1998, Ambulatory electrocardiographs, Arlingon (VA): Association for the Advancement of Medical Instrumentation, 1999.

[Di 97] J. Di, *Hochpräzise "Switched Current" Schaltungen*, Ph.D. thesis, ETH Zurich, 1997.

[Clip 93] A. J. Clippingdale, *The sensing of spatial electrical potential*, Ph.D. thesis, University of Sussex, 1993.

[IEC 08] International Electrotechnical Commission standard IEC/EN 60601-1, Medical electrical equipment, International Standards and conformity assessment for government, business and society for all electrical, electronic and related technologies, 2008.

[Huij 96] J. H. Huijsing, *Design of low-voltage low-power operational amplifier cells*, Springer, 1996

[Hwan 79] K. Hwang, *Computer Arithmetic*, John Wiley and Sons, 1979.

[Guyt 00] A. C. Guyton and J. E. Hall, *Textbook of Medical Physiology*, W.B. Saunders Company, 2000.

[Jesp 01] P.G.A. Jespers, *Integrated Converters*, Oxford University Press, 2001.

[Menk 89] W. Menke, *Handbuch Medizintechnik*, Ecomed, Landsberg/Lech, 4. Auflage, 4 Ordner, 1989-...

[Mett 93] A. C. Metting van Rijn, *The Modelling of biopotential Recordings and its Implications for Instrumentation Design*, Ph.D. thesis, Technische Universität Delft, 1993.

[Nobe 65]	*Nobel Lectures, Physiology or Medicine 1922-1941*, Elsevier Publishing Company, Amsterdam, 1965.
[Plas 94]	R. van de Plassche, *Integrated Analog-to-Digital and Digital-to-Analog Converters*, Kluwer Academic Publishers, 1994.
[Tiet 02]	U. Tietze and Ch. Schenk, *Halbleiter-Schaltungstechnik*, Springer-Verlag, 12th edition, 2002.
[Webs 98]	J. G. Webster, Editor, *Medical Instrumentation, Application and Design* John Wiley and Sons, 3rd edition, 1998.
[Wick 04]	R. Wicki, *Herzschlag-Detektion mit Radar - Eine Machbarkeitsstudie* diploma thesis, Institute of Scientific Computing, ETH Zurich, 2004.
[Wies 01]	H.G. Wieser, "Foramen ovale electrodes" in *Epilepsy Surgery, pp. 573–584.* Lüders HO, Comair YG (eds), Philadelphia, scnd ed., 2001.
[Zsch 02]	S. Zschocke, *Klinische Elektroenzephalographie* Springer-Verlag, Heidelberg, 2.Aufl., 2002.

Products

[bios]	"active one," http://www.biosemi.com, last accessed Nov. 2004.
[vita]	"vitaphone," http://www.vitaphone.de, last accessed Nov. 2004.
[cadi]	"Cadiscope," http://www.strela-development.net, last accessed Nov. 2004.
[hso]	"HSO-Link," http://www.strela-development.net, last accessed Nov. 2004
[stre]	"Strela Development AG," Sennweidstrasse 45 CH-6312 Steinhausen
[uniSG]	"Cumulative Thesis at the University of St. Gallen," http://www.studium.unisg.ch/Studium/Doktorat/AllgemeineInformationen/Dissertationsformen.aspx, last accessed March 2011

Projects

[cti1]	H. Jäckel, and F. Voegeli, "Aktive Biosignal-Sonden (für EEG, EKG etc) mit drahtloser Daten- und Energieübertragung," *KTI-Projekt 4431.1*

[cti2]   M. Weder, and U. Froitzheim, "Personal Textile Electrodes PTE," *KTI-Projekt 7819.2*

[empa]   "Ein intelligentes T-Shirt überwacht das Herz" *Jahresbericht 2006*, pp. 32–33, *http://e-collection.ethbib.ethz.ch/ecolpool/journal/empa_jahresbericht/2006.pdf*, last accessed Dez. 2007.

[EU1]   EMERGE "Emergency Monitoring and Prevention," *EU-project 045056*

Reports

[degen]   T. Degen, "Akustische Messungen an elektronischen Stethoskopen," interner Bericht, IfE, ETHZ *http://www.library.ethz.ch*

[wyru]   S. Wyss, and M. Rufer, "Wearable lifesaver II" Diploma Thesis, IfE, ETHZ, ws2002/2003

[maku]   C. Mattman, and A. Kuhn, "Truly Wireless ECG," Semester Thesis, IfE, ETHZ, ss2003

Patent Applications

[USpat]   T. Degen, "Pseudo differential amplifier with DC-offset compensation using two-wired preamplified active electrodes" Patent Application, US60/604,513, 27. December 2004

ACKNOWLEDGMENTS

I would like to thank all of you who have in some way or another contributed to this thesis. Most of all I would like to thank Prof. Dr. Hans-Andrea Loeliger who stepped in when my former professor decided to abandon my thesis. Thanks are also going to Dr. Vetter and Dr. Schmid for accepting to be co-examiners. Especially Dr. Schmid invested an incredible amount of time in proofreading the thesis and it was also him encouraging me to reduce the number of pages to a mere hundred.

Consequently, I would like to express my gratitude to all members of the electronics laboratory for their support. This is especially true for Theodor, Markus and Matthias, who were true friends during our common time in the office. I would like to thank all three for showing me new angles to look at the world and the meaning of 'Parabelflug' (zero-g flight) including means to qualify it.

My special thanks go to Ruth, Eveline and Esther for organizing the daily life. To Fritz for his moral support and for the whole analog group being there whenever needed. To the digital group for all the coffee and the student projects in common and of course to the students always eager to work on my projects. Without Ruth I would have abandoned this work before the end.

Finally, this thesis would never have been possible without the continuous backing of Christina, my beloved wife, thank you so much!

Special Thanksgiving

I wish to thank Christina, Mr. Fridolin Voegeli, Mr. Michael Breuss, Ms. Veronica Ossevoort, Dr. Julian Randall and Ms. Michelle Jenkins for proof-reading the different manuscripts. Special thanks to Mr. Jhonny Lloyd and Ms. Susi Trenka for proof-reading my publications.

Additional Acknowledgments

I will also acknowledge that this thesis took far too long and that I am partially responsible for it. On the other hand, it is a pity that there are no official rules for the acceptance of a cumulative thesis even though the use of previously published material is, under few conditions, explicitly permitted by the implementation rules of the rector of the ETH Zurich. I was shocked to realize that these rules are not respected and I would even argue that this is not to the advantage of the ETH Zurich nor does it benefit our economy. It certainly does not help to reduce the timespan required for a Ph.D.

Cumulative dissertations are not something subversive, Einstein did one at the university of Zurich. The university of St. Gallen has very clear rules regarding the subject: "A cumulative thesis must consist of at least three essays reflecting the quality standard required by reputed international journals; they may already have been published. At least one essay must have been contributed by the Ph.D. student him or herself and must comprise a substantial part of the thesis as a whole"[uniSG]. I would wish the rules at the ETH Zurich would be as clear as the rules at the university of St. Gallen.

At some universities it is common practice to conclude the thesis with a set of theses. I follow this tradition by naming five theses for a successful Ph.D. program which removes part of the arbitrariness.

Five Theses for Future Successful Ph.D. Thesis

1. A Ph.D. is a contract between a young researcher and the *department* of an institute for higher education like the ETH Zurich.

2. The department appoints the thesis *committee* consisting of a professor being appointed as advisor to the thesis, at least one co-examiner and the chair of the committee (a member of the department) before the beginning of the thesis.

3. The *professor* defines the content and the goals of the Ph.D.

4. After one year the Ph.D. candidate presents his research plan to the committee for approval. After this approval the content of the research plan cannot be changed without written consent of the thesis committee.

5. Every year the state of the thesis work is presented to the thesis committee until the thesis is finished, successfully or not.

INDEX

List of abbreviations

V_{CM} *(common-mode voltage)*, 15
V_{DM} *(differential mode voltage)*, 16
V_{IM} *(isolation mode voltage)*, 16
V_{PO} *(body voltage)*, 16
AAMI *(association for the advancement of medical instrumentation)*, 14
ADC *(analog to digital converter)*, 100
CMR *(common mode range)*, 55, 68
CMRR *(common mode rejection ratio)*, 18
DA *(differential amplifier)*, 26
DAC *(digital to analog converter)*, 103
DRL *(driven right leg)*, 34
EMPA *(Swiss Federal Laboratories for Materials Science and Technology)*, 121
FET *(field effect transistor)*, 78
FFT *(fast Fourier transform)*, 62, 83
FIR *(finite impulse response)*, 46
GBP *(gain-bandwidth-product)*, 97
IEC *(International Electrotechnical Commission)*, 14
INA *(instrumentation amplifier)*, 26
op-amp *(operational amplifier)*, 13, 90
PCB *(printed circuit board)*, 62
USZ *(university hospital, Zurich)*, 62

i want morebooks!

Buy your books fast and straightforward online - at one of world's fastest growing online book stores! Environmentally sound due to Print-on-Demand technologies.

Buy your books online at
www.get-morebooks.com

Kaufen Sie Ihre Bücher schnell und unkompliziert online – auf einer der am schnellsten wachsenden Buchhandelsplattformen weltweit! Dank Print-On-Demand umwelt- und ressourcenschonend produziert.

Bücher schneller online kaufen
www.morebooks.de

VDM Verlagsservicegesellschaft mbH
Heinrich-Böcking-Str. 6-8   Telefon: +49 681 3720 174   info@vdm-vsg.de
D - 66121 Saarbrücken   Telefax: +49 681 3720 1749   www.vdm-vsg.de

Printed by Books on Demand GmbH, Norderstedt / Germany